GREEN PRODUCTS

Perspectives on Innovation and Adoption

GREEN PRODUCTS

Perspectives on Innovation and Adoption

Edited by
João Neiva de Figueiredo
Mauro F. Guillén

CRC Press
Taylor & Francis Group
Boca Raton London New York

CRC Press is an imprint of the
Taylor & Francis Group, an **informa** business

A PRODUCTIVITY PRESS BOOK

CRC Press
Taylor & Francis Group
6000 Broken Sound Parkway NW, Suite 300
Boca Raton, FL 33487-2742

© 2012 by Taylor & Francis Group, LLC
CRC Press is an imprint of Taylor & Francis Group, an Informa business

No claim to original U.S. Government works

Printed in the United States of America on acid-free paper
Version Date: 20110803

International Standard Book Number: 978-1-4398-5465-5 (Hardback)

Library of Congress Cataloging-in-Publication Data

Green products : perspectives on innovation and adoption / editors, João Neiva de
Figueiredo, Mauro F. Guillén.
    p. cm.
Includes bibliographical references and index.
ISBN 978-1-4398-5465-5 (hardcover : alk. paper)
    1. Green products. 2. Energy conservation. 3. Transportation--Environmental
aspects. I. Figueiredo, João Neiva de. II. Guillén, Mauro F. III. Title.

HD9999.G772G744 2012
333.79'4--dc23   2011030274

Visit the Taylor & Francis Web site at
http://www.taylorandfrancis.com

and the CRC Press Web site at
http://www.crcpress.com

Contents

Foreword vii
Editors ix
Contributors...xi

Chapter 1 Mapping the Universe of Green Products....................... 1
  João Neiva de Figueiredo, Mauro Guillén, and Xiaoting Zheng

Chapter 2 The Need for Sustainable Energy Systems 17
  José Gómez Moreno

Chapter 3 The Transition from Oil Dependency to
  Sustainability in Denmark 35
  José Gómez Moreno

Chapter 4 The Revival of Battery-Powered Electric Vehicles in
  Japan.. 49
  Jan Zelezny

Chapter 5 Sustainable Urban Mass Transport................................ 81
  Farheen Qadir

Chapter 6 The Promise of Sugarcane Ethanol as a Cleaner
  Combustion Engine Fuel 107
  Jeffrey B. Bryant and María Teresa Burbano

Chapter 7 The Challenge of Sustainable Tourism......................... 135
  Jessica Webster

Chapter 8 Conclusion: Green Product Innovation and Adoption 165
  João Neiva de Figueiredo and Mauro Guillén

Bibliography 173

Appendix 1: The Green Products Universe 187

Appendix 2: Essential Readings on Green Products 189

Index 203

Foreword

Green products and services have gained in acceptance and use throughout the world. This book represents an effort to map this growing universe and to offer specific examples from several countries around the world. It is the result of the joint effort of researchers affiliated with the Joseph H. Lauder Institute of Management & International Studies and the Wharton School at the University of Pennsylvania. Funding for this project came primarily from the Leon Lowenstein Foundation. We are grateful to Thomas Bendheim, member of the Lauder Institute Board of Governors, for his unwavering support. We are also indebted to Kim Norton, Julia Zheng, and Xiaoting Zheng, who helped the research team throughout the various stages of the process.

João Neiva de Figueiredo and Mauro F. Guillén
Philadelphia

Editors

João Neiva de Figueiredo is Assistant Professor in the Department of Management at Saint Joseph's University's Haub School of Business and Senior Fellow at the University of Pennsylvania's Wharton School. With over 20 years experience in international business, Dr. Neiva was a consultant at McKinsey & Company, a vice president at Goldman Sachs, and a partner at JPMorgan Partners. His research and teaching focus on the role and effects of sustainability practices in the areas of strategy and international management. Dr. Neiva holds Electrical and Systems Engineering degrees from Rio de Janeiro's PUC, an MBA with high distinction (Baker Scholar) from the Harvard Business School, and a PhD in Business Economics from Harvard University.

Mauro F. Guillén is Director of the University of Pennsylvania's Joseph H. Lauder Institute of Management & International Studies, and Dr. Felix Zandman Professor of International Management at the Wharton School. He has written extensively about multinational firms, economic development, and the diffusion of innovations. He received his PhD in Sociology from Yale and a doctorate in Political Economy from the University of Oviedo in his native Spain.

Contributors

Jeffrey B. Bryant is a member of The University of Pennsylvania undergraduate class of 2010.

María Teresa Burbano is a member of The University of Pennsylvania undergraduate class of 2010.

José Gomez earned an MBA with honors from The Wharton School (The University of Pennsylvania) and MA in international studies from the Lauder Institute of The University of Pennsylvania's School of Arts & Sciences.

Farheen Qadir graduated from The University of Pennsylvania's Lauder Institute joint MBA/MA program in 2010 and is with BASF in Iselin, New Jersey.

Jessica Webster graduated from The University of Pennsylvania's Lauder Institute joint MBA/MA program in 2010 and is with Expedia.

Jan Zelezny graduated from The University of Pennsylvania's Lauder Institute joint MBA/MA program in 2010 and is with Citigroup's Investment Banking Division in Tokyo.

Julia Zheng is a member of The University of Pennsylvania undergraduate class of 2012.

Xiaoting Zheng is a member of The University of Pennsylvania undergraduate class of 2012.

1

Mapping the Universe of Green Products

João Neiva de Figueiredo, Mauro Guillén, and Xiaoting Zheng

CONTENTS

1.1   Green Energy Sources 2
1.2   Green Energy Use 4
1.3   New Environmental and Recyclable Products and Materials 6
1.4   Green Water Management...7
1.5   Lifestyles and Green Consumption...9
1.6   Green Policies and Regulation 10
1.7   Adapting Business Practices to Green Production 11
1.8   Five Case Studies of Green Production and Consumption 12
        1.8.1   Energy Systems..13
        1.8.2   Electric Engines for Automobiles 13
        1.8.3   Urban Mass Transit 14
        1.8.4   Ethanol Fuels 14
        1.8.5   Green Tourism..15

Research, innovation, and commercialization of green products have taken a quantum leap over the past decade. Each year dozens of path-breaking innovations are commercialized, ranging from energy-efficient lighting to recyclable packaging, and from water filters to organic fertilizers. Growing concerns over the impact of human activity on the natural environment have made green products the subject of much analysis and debate (Chung and Tsai 2007). A recent McKinsey survey conducted in developed and emerging economies showed that nearly 90% of consumers worry about the environmental and social impact of the goods and services they buy, although no more than one in three is specifically willing to buy green products. The most popular green products among consumers are efficient lighting and organic foods. By contrast, few consumers buy green automobiles or detergents unless they also fully satisfy other desired attributes (Bonini and Oppenheim 2008).

There is no precise and generally agreed-upon definition of what a green product is. One common definition is that a "green product" is a good or service designed to minimize its environmental impact over the entire product life cycle (Albino, Balice, and Dangelico 2009). The term "green product" is broadly used to describe products that integrate environmental requirements within the initial design process, including minimizing raw materials and energy consumption, waste generation, health and safety risk, and ecological degradation (Baumann, Boons, and Bragd 2002). The design and development of green products, also called eco-design or design for the environment, is a product development process that takes into account the complete life cycle of a product and considers environmental aspects at all stages, striving for products that have the lowest possible environmental impact throughout their entire life cycle (Glavic and Lukman 2007). In this book, we pay attention to **green production and consumption** in encompassing terms, including all of the activities necessary to offer green goods and services, that is, research and development (R&D), production, distribution, marketing, sales, and servicing.

The universe of green products is a huge landscape. Appendix 1 offers a classification of green goods and services, including clean and renewable energy sources, energy use, new environmental and recyclable products and materials, water management, and lifestyle-related green products. In addition, we also map environmental and sustainability policies and business practices that contribute to the development and diffusion of green products.

1.1 GREEN ENERGY SOURCES

Perhaps the most rapidly growing area of green production has to do with **clean and renewable energy sources**. Global warming has prompted much (subsidized) investment in this key area, under the assumption that technology takes time to develop and that eventually costs can be brought down through innovation and learning curve effects (Reddy and Painuly 2004). Renewable sources of energy presently account for about 3% of total energy use and about 18% of total electricity generation (Lior 2008). Renewable energy has the potential to supply a substantially larger part of the world's anticipated energy needs. However, there still exist significant barriers preventing the adoption of renewable energy sourcing. These need to be addressed before widespread use of renewable energy sources becomes commonplace. Certain barriers include technical limitations, lack of awareness and information, economic constraints, regulatory barriers, market failures, and consumer behavior (Reddy and Painuly 2004).

Among all renewable energy sources, **solar energy** in the form of solar electromagnetic radiation is the most ancient and in fact the original source of almost all fossil and renewable energy types. The total solar radiation power received on the Earth's surface is 1.73×10^{14} kW every year, approximately 10,000 times greater than total human energy consumption (Şen 2004). Systems that capture solar energy and convert it into electricity are among the most environmentally friendly ways to meet our energy needs (Kalogirou 2004). So far, solar energy is only a minor contributor due to cost and intermittency constraints, but given many recent technical improvements, experts predict that it will become a major component of electricity production in the near future (Fthenakis, Mason, and Zweibel 2009). Although recent growth has flattened because negative environmental impacts can be massive (Sarkar and Karagoz 1995), **hydroelectric power** generation accounts for nearly 90% of global renewable energy production (Glasnovic and Margeta 2009). Small hydropower capacity has increased to an estimated 85 GW worldwide and large hydropower to 860 GW as of the end of 2008 (REN21 2009).

Most of the recent growth in renewable energy production has come from the **wind power** sector, primarily driven by government subsidies (Karki 2007). Between 1999 and 2009, growth in cumulative installed capacity has been approximately 27% per annum (*Global Wind* 2009). Wind power is regarded as the most promising means of achieving a significant reduction in fuel use and greenhouse gas emission in the near future (Karki 2007).

Geothermal energy, or the natural heat of the earth transferred from the interior to the surface by conduction, has also seen rapid growth (Fridleifsson 1996). Over the past 30 years, worldwide installation of geothermal capacity increased from 1.3 GW in 1975 to 10.5 GW in 2009 (Earth Policy Institute 2010). This growth was made possible by a number of technical advances in both electricity generation and direct applications. Recently, however, economic cycles, a decrease in government incentives, price fluctuations in fossil fuels, and increasing attention to other renewable energy sources, such as wind, have discouraged geothermal power development (Gallup 2009).

Biomass production represents another promising area. It is defined as all organic materials produced by plants through photosynthesis, where the energy of sunlight is stored in chemical bonds and released when the bonds break. The three main types of biomass conversion product are electrical/heat energy, transport fuel, and chemical feedstock (McKendry 2002). Biomass has always been a major energy source, with a total power generation capacity of 52 GW and total heat capacity of about 250 GW as of the end of 2008 (REN21 2009). The increasing utilization of biomass energy has the potential to offset the use of fossil fuels and reduce global warming. However, it is also a controversial alternative given the threat that it poses to conservation areas, water resources, and food safety. In addition, effects on carbon balance and land use need to be taken into account (Field, Campbell, and Lobell 2008).

Perhaps the most controversial green energy source is **ethanol**, which is produced through the fermentation of agricultural products, mainly sugarcane in Brazil and corn in the United States (see Chapter 6). Ethanol fuel can achieve a positive net effect on global warming (Goldemberg, Coelho, and Guardabassi 2008). As of the end of 2008, total world capacity of ethanol production stood at 67 billion liters per year, equivalent to a 5.4% share in global gasoline-type fuel use (REN21 2009). As with other green energy sources, there are drawbacks: ethanol in gasoline may result in lower urban air quality and risks to water resources and biodiversity (von Blottnitz and Curran 2007).

1.2 GREEN ENERGY USE

Energy consumption is involved in every activity of daily life. Under the threat of the ever-increasing global energy demand, how to use the limited energy sources more efficiently has become a critical issue. Energy efficiency is defined by Horace Herring, a professor at The Open University in the UK, as the ratio of energy services output to energy input (Herring 2006). Over the past 20 years, various new technologies and organizational arrangements have brought about significant increases in energy efficiency, especially in the areas of mass transit solutions, alternative automobile engines, building designs, new manufacturing processes, lighting, and household appliances.

Along with the booming global economy and the ongoing rural exodus in developing economies, the demand for **mass transit** has increased significantly over the past few years, with accompanying growth in energy consumption and carbon dioxide emissions (Roth and Kåberger 2002). There are many potential solutions to the energy conservation and overall sustainability of transportation systems, including increased fuel efficiency, alternative fuel supplies, and improved vehicle design. In this area, other considerations such as congestion reduction, land use, and traffic safety are also important (Litman 2005).

In response to its critics, and as a way of developing a sustainable future for itself, the automobile industry has long been active in fuel-saving technologies and **green vehicle** designs (Schäfer and Jacoby 2006). Currently the main focus of research, development, and commercialization is on battery-powered electric vehicles, hybrid electric vehicles, and fuel cell electric vehicles, together with internal combustion engine alternatives such as compressed natural gas vehicles and ethanol- and methanol-fueled vehicles (Åhman 2006; see Chapter 4).

Perhaps the area with the greatest potential is **green buildings**, defined as those with minimal adverse impact on the environment, including the buildings themselves, their immediate surroundings, and the broader urban, regional, and global settings. Five oft-cited objectives for sustainable buildings are resource efficiency, energy efficiency, pollution prevention, harmonization with the environment, and integrated and systemic approaches (John, Clements-Croome, and Jeronimidis 2005). Many practices can contribute to making buildings greener, for example utilizing renewable energy and environmentally friendly building materials.

Globally the manufacturing sector accounts for about 40% of primary energy use (Price, Worrell, Sinton, and Yun 2001). Thus, **green manufacturing processes** hold great promise as a way to attain sustainable economic growth while preserving the environment. Increasing industrial energy efficiency will not only reduce the use of energy but also bring additional productivity benefits, such as lower maintenance costs, increased production yield, and safer working conditions (Ayres, Turton, and Casten 2007).

Green lighting is widely regarded as an important potential contributor to energy efficiency and environmental protection. Actions in this area span the entire value chain and life cycle of lighting products, including design, installation, and use (Loe 2003). However, as important as energy efficiency is, it must be achieved without endangering the overall quality of the lighting to ensure productive and safe home, workplace, and transportation environments.

Finally, **green household appliances** are a key area because such devices account for about 20% of total energy consumption (Dzioubinski and Chipman 1999). The use of appliances in the home shapes household electricity requirements, a form of direct energy use, and involves indirect energy use as well, such as production, transportation, and disposal (Steg 2008). To develop and promote energy-efficient household appliances is an obvious way to improve household energy conservation, but possible side effects such as safety considerations also need to be taken into account (Steg 2008).

1.3 NEW ENVIRONMENTAL AND RECYCLABLE PRODUCTS AND MATERIALS

Durable and non-durable consumer products have a huge negative impact on the environment. Urban wastes with poisonous chemical compounds, heavy metal elements, and non-biodegradable materials have contaminated significant portions of the soil, water, atmosphere, and the whole ecosystem. In order to reduce this impact and establish a more sustainable system, efforts to recycle and reuse waste and to develop more environmentally friendly materials and products need to be intensified.

Green recycling practices have the potential of reducing environmental damage in significant and far-reaching ways. For example, at the present time about 80 to 90% of end-of-life electronic products are being sent to landfill sites. Design for Recycling (DFR) technologies could change this situation (SRL 2009). As defined by the Institute of Scrap Recycling Industries, DFR seeks to achieve two very basic goals: first, to eliminate or reduce the use of hazardous or toxic materials that may present a grave danger to the environment or put recycling workers in jeopardy; and second, to discourage the use of materials that are not recyclable or manufacturing techniques that make a product non-recyclable using current technologies (ISRI 2009). The adoption of the DFR concept in the product design process is critical to address the waste associated with electrical and electronic equipment (WEEE), and end-of-life vehicle (ELV).

An exciting area of research, development, and commercialization is **green plastics** or bioplastics. A biopolymer is a special polymer that involves living organisms in its synthesis process (ASTM Standard D6866-06). Bioplastics therefore are defined as materials that contain biopolymers in various percentages and can be shaped by heat and pressure, and they are thus considered potential alternatives to conventional thermoplastic polymers of petrochemical origin (Queiroz and Collares-Queiroz 2009). Although the bioplastics industry is still in its infancy, it has made enormous progress and a strong expansion is expected in the near future.

Green packaging also harbors great potential, given that virtually all consumer products and many services require containers to protect, preserve, transport, or use them. Packaging waste accounted for 78.8 million tons or nearly 32% of total municipal solid waste (MSW) in 2003 in the United States. The dominant disposal method is still landfill, which occupies valuable space and creates adverse environmental effects (Kale et al. 2007). Promising solutions to reduce MSW include recycling the commonly used packaging materials, such as steel, aluminum, paper, and plastic, and composting some of the conventional materials with advanced materials, such as biopolymers.

Besides green recycling and packaging, **green waste management** can make important contributions to sustainability. New techniques and practices can help avoid water, soil, and air contamination. Some of the most common waste management practices include landfill, incineration, recycling, and composting, and the actual adoption varies among different countries (Giusti 2009). Currently, about 20% of MSW is recycled in the United States, Japan, the UK, and France (Giusti 2009). More cost-effective and environmentally friendly technologies are needed to alleviate the impact of waste on the environment (Nowosielski and Zajdel 2007).

1.4 GREEN WATER MANAGEMENT

Households, factories, and farms depend on a reliable and safe supply of water. Of worldwide human water consumption 70% is for irrigation purposes, while industrial usage accounts for 20%, and household usage accounts for 10%. Fortunately, since the 1960s conservation efforts and new technologies have reduced consumption from between 0.3 and 0.4 cubic meters per dollar of gross domestic product (GDP) to less than 0.1

in most developed countries. Developing and emerging economies have also reduced water use to comparable levels (UNESCO 2009, 88, 99, 108, 109, 111).

Although **green agriculture** was proposed several decades ago, experts argue that a methodological base to systemically assess agricultural sustainability has not yet been developed, and there are no systematic operational practices to implement sustainability on a large scale (Sydorovych and Wossink 2008). Agricultural sustainability includes economic, social, and ecological aspects, of which water management is a significant element. Irrigation, in particular, can more than double land productivity and economic returns, but it can also have unwanted environmental consequences when poorly managed, including water logging and soil salinity (Khan et al. 2006).

Green gardening and landscaping represents yet another important area because of its great social significance and potential for recreational, economic, and human health benefits (Syme et al. 2004). The amount of water used in gardens accounts for more than half of total household water usage. Fortunately, people tend to be much more price-sensitive about using water for gardening than for indoor use (Syme et al. 2004). Better landscape water management requires well-designed irrigation systems, a watering schedule sensitive to changing conditions, and continuous monitoring to achieve a more efficient use of water resources (Irrisoft 2009).

Due to the increasing presence of new compounds in wastewater brought by new developments in various industries, it is vital to research and develop new or more efficient **green water treatment** technologies (Gogate and Pandit 2004). The conventional wastewater treatment techniques are classified into centralized systems and decentralized systems, which may use non-biological, biological, and ecological methods (Burkhard, Deletic, and Craig 2000). New alternatives are focused mainly on oxidation technologies at ambient conditions and hybrid methods (Gogate and Pandit 2004). Creating a more efficient sewage system is a complex process, which requires both thoughtful planning and social acceptance and participation (Burkhard et al. 2000).

Water purification has emerged as a necessary technology given contamination and scarcity in certain parts of the world (Shannon et al. 2008). Conventional methods of water purification are very effective, but they are also chemically, energetically, and operationally intensive, and require large-scale implementation in order to be cost efficient. New sustainable and efficient technologies for water disinfection and decontamination are being developed at the present time, but much work lies ahead.

Natural freshwater accounts for less than 0.5% of the total amount of water on Earth, and much of it is stored underneath the surface and cannot be accessed economically (Khawaji, Kutubkhanah, and Wie 2008). **Desalination** has thus become a necessity in certain parts of the world (von Medeazza and Moreau 2007). Over the past years, the daily production of desalination plants operating worldwide has increased from 13 million m^3 to 32 million m^3, supplying 160 million people (Schiffler 2004). The most important technologies are multi-stage flash distillation (MSF), multi-effect distillation (MED), and reserve osmosis process (RO) (Khawaji et al. 2008).

1.5 LIFESTYLES AND GREEN CONSUMPTION

Human beings and the surrounding environment are continuously interacting with each other. The environment is impacted by us in a number of ways, and at the same time the environmental consequences of this impact affect people's preferences, choices, and decisions in daily life. Food consumption patterns, sustainable tourism, and waste disposal habits are three important areas of research and implementation.

The environmental risks associated with **food consumption patterns** are large and complex (Halkier 2001). Consumers have become more concerned with food quality not only in terms of nutritional content but also in terms of occurrence of food additives, agrochemical refuse, and contamination from environmental pollution (Wandel and Bugge 1997). Research has unveiled food product issues that inevitably affect consumers' decisions and habits, including additives in manufactured goods, pesticide residues, chemicals, growth hormones, and genetically modified organisms (Halkier 2001). Moreover, the inefficiencies and impact on global warming of beef production have been noted.

Another important lifestyle issue has to do with **green tourism**. Despite the fact that tourism has become one of the world's largest industries and has brought huge economic benefits, it has also significantly contributed to environmental, social, and cultural damage (Choi and Sirakaya 2006). The World Tourism Organization estimates that travel for leisure accounts for 5% of carbon emissions (Scott et al. 2010). Tourism planners and operators have recently begun to address these issues on more than a symbolic scale (Yaw 2005). Green tourism cannot become more widespread unless ecological, social, economic, political, cultural, and technological dimensions are carefully researched (Choi and Sirakaya 2006). New practices need to be designed and implemented (Tepelus 2005).

With the growth of human population, recent urbanization, and industrialization, waste has become a serious issue worldwide. Waste is generated in all kinds of human activities, with an amount estimated to exceed 2 billion tons per year globally (Key Note 2007). **Greener practices in waste generation, recycling, and disposal** have the potential for bringing about huge improvements in the quality of life. The cost efficiency of urban waste management, for instance, depends on various kinds of techniques, including the technology used, the popularity and prevalence of waste collection, and service delivery (Lombrano 2009).

1.6 GREEN POLICIES AND REGULATION

The past two decades have witnessed an explosion in government intervention to protect the environment and promote sustainable development. Several policymaking levers have been used with varying degrees of success, including regulation, standard setting, diffusion, and subsidies.

Although **green regulations** are an essential component to protect the environment in many areas, actual enforcement and compliance are critical but often overlooked (Heyes 2000). The effects of environmental regulations on industrial performance, economic growth, inequality, and the efficient allocation of resources are very much under debate (Qi, Zheng, and Zhao 2007).

Green standards, whether voluntary or not, have become a major way of encouraging environmentally friendly production and consumption. Pressure from governments, non-governmental organizations, and consumers have driven the construction of international environmental and social standards. The most impactful to date are the ISO standards (Schmitz 2008).

Another area of research and implementation involves the **diffusion of green practices**. Market and non-market mechanisms are both important to this process of widespread adoption (OECD 2005). It is especially important to gain a better understanding of the diffusion process in terms of feedback mechanisms and multi-directional linkages (Montalvo and Kemp 2008). Factors found to affect the adoption of environmentally responsible practices include public policy, economic benefits and risks, market feedback, social pressure, public attitudes, technologies, and organizational capabilities (Montalvo 2008).

Government **subsidies**, or negative taxes, are a way to encourage green production and consumption practices and reduce adverse environmental impact. Subsidies have different forms and purposes, including those aimed at the design and production of environmentally friendly goods and services, the promotion of renewable energy sources, and the reduction of material use and waste, to name but a few. Although recent research suggests that subsidies can result in an increase in sales of green products (Olsson and Gärling 2008), can help reduce greenhouse gas emission, and can promote firms' investment in green R&D (Aalbers et al. 2009), what continues to be a subject of heated debate is whether or not they are the most effective and efficient way to achieve those goals.

1.7 ADAPTING BUSINESS PRACTICES TO GREEN PRODUCTION

The green product revolution creates new opportunities and challenges to businesses. Most need to adapt their practices to new technologies, changing government policies, and shifting social demands and expectations. Such adaptations are complex, as they involve different layers of management, organization, communication, and cooperation.

The past few decades have seen a growing interest in **researching and developing green products**. Improvements made to goods and services so as to reduce negative environmental impacts may result, on the one hand, in higher development and production costs, and, on the other, in higher product differentiation, increased margins, and larger market shares for those firms which successfully move to commercialize green products (Reinhardt 2008). Only a very small fraction of businesses around the world have adopted systematic approaches to R&D for green products (Baumann et al. 2002).

As environmental issues and concerns take on an increasingly important role, conventional **accounting** practices are no longer sufficient to provide adequate information for decision-making on environmental issues at firms (Jasch 2006). It is of utmost importance to identify both the visible and the hidden costs and profits related to the environment (Joshi, Krishman, and Lave 2006). The first step toward fair pricing is to identify and quantify hidden environmental and social costs that are part of companies' value chains. During recent years, environmental management accounting (EMA) has been proposed and promoted by organizations in different countries (Jasch 2006).

Beginning in the early 1990s, research showed that consumers yearned for green goods and services. Thus, **green marketing** emerged as a new subfield. However, consumers are less willing to buy green products than their attitudes toward the environment might suggest. Green products still account for small market shares across various product categories (Rex and Baumann 2007). The most important inhibiting factors that green marketing needs to overcome are lack of awareness, lack of trust, and unwillingness to pay higher prices for environmentally friendly products (Bonini and Oppenheim 2008). Companies will need to develop marketing strategies aimed at educating consumers, improving product awareness, and popularizing green products.

Finally, green production and distribution may require changes in the **organization** of the firm. A green business firm has been defined as "an organization that behaves in a manner that takes into account the long-term effects of its activities on the natural environment and does not attempt to externalize the costs of those activities to other stakeholders or the environment itself" (Jones 1996, 328). In order to take advantage of environmental regulations and standards, internal and external pressures, and technology advancements, new environmental strategies need to be developed and adopted by firms and organizations. One innovative idea in this respect is environmental management systems (Albino et al. 2009).

1.8 FIVE CASE STUDIES OF GREEN PRODUCTION AND CONSUMPTION

This book includes five in-depth case studies of innovative green goods and services, shedding light on the technological options available, the opportunities for commercialization, the organizational and managerial requirements, and conditions for widespread adoption of green technologies.

1.8.1 Energy Systems

In the past few years, discussions regarding the need and means to implement sustainable policies in the energy sector have become widespread. Several arguments, such as the expected medium-term mismatch between energy demand and supply, the uncertainty of supply sources, and the behavior of prices, support the thesis that the energy sector would benefit from significant changes in order to ensure its long-term sustainability. These arguments, together with climate change, are shifting the perceptions of governments, businesses, and consumers toward sustainability, but the lack of the right incentive system is preventing the transformation of this awareness into meaningful action. Sustainable changes in the energy system should revolve around increasing energy efficiency in the generation, distribution, and use of energy; upgrading the electricity grid; and increasing use of renewable resources. Chapter 2 offers an overview of these trends.

Some countries have already started to implement sustainable changes in their respective energy sectors with varying levels of success. Denmark was one of the first countries to implement such changes and therefore constitutes a representative case of the conditions and policies necessary to achieve them. Denmark was almost totally dependent on foreign fossil resources in the early 1970s and, over the past four decades, has transformed its energy system into one of the most sustainable in the world while maintaining its economic prowess. Chapter 3 analyzes the most significant factors, events, and policies that made Denmark a role model in the transition to a sustainable energy system.

1.8.2 Electric Engines for Automobiles

The cleaner automobile development effort encompasses many approaches that can be divided into four basic technologies: biofuel-based automobiles; flex-fuel-based automobiles with engines that run on any combination of ethanol and gas; hybrid and plug-in hybrid automobiles that have both electric and combustion engines; and pure electric engine-powered automobiles. The latter are zero-emission vehicles based on all-electric power-trains, representing the cleanest of the alternatives listed (although there still are unresolved environmental issues associated to battery disposal). Chapter 4 examines the current stage of development of electric cars, compares the different electric engines in initial stages of adoption, and investigates selected companies' expected upcoming market offerings in Japan.

1.8.3 Urban Mass Transit

Urban mass transit systems that are both environmentally and socially sustainable are becoming essential to managing the continued growth of large cities in developing countries. In Mexico City, one of the three largest cities in the world, car use has doubled between 2002 and 2009, and it is estimated that every year, for every baby born, 1.6 new cars enter the city's fleet. Moreover, the city's air pollution levels exceeded acceptable limits on 60% of the days of the year. To address this issue of air pollution and increased traffic, a unique urban mass transit solution, the Metrobús, has been implemented in Mexico City, carrying 320,000 passengers per day by the end of its first year of operation. As a result of its success, it is currently being expanded to several other large Mexican cities. Apart from Mexico, Latin American implementations of this type of public transportation system include the Rede Integrada de Transporte in Curitiba (Brazil); Transmilenio, in Bogotá (Colombia); Trolebús, in Quito (Ecuador); and Transantiago, in Santiago (Chile). In 2007 Istanbul began implementation of such a project, which currently transports over 500,000 passengers per day. Although this type of system was first introduced over 30 years ago in Curitiba, only in recent years has it become widespread. The case of Mexico City is important because it represents the successful implementation of the system in a fast-growing megalopolis. Chapter 5 examines the implementation of the Metrobús in Mexico City and seeks to understand the requirements for success of such a system, analyzing its potential for success in other developing nation urban centers with explosive growth, such as Cairo, Egypt, for example.

1.8.4 Ethanol Fuels

Sugarcane ethanol has become a viable and efficient energy source for gasoline blends. Whereas corn ethanol obtains 1.3 to 1.5 units of energy output for every unit of fossil energy input used in its transformation, the same ratio for sugarcane ethanol is eight to one, making this process several times more efficient. The stalk of the sugarcane is 20% sugar, which is fermented to make alcohol, and the waste cane can be burned to help power the transformation process leading to usage of lower quantities of fossil fuel. Furthermore, sugarcane yields 600 to 800 gallons of ethanol per acre, twice as much as corn (per acre). Brazil has pioneered the large-scale production and distribution of sugarcane ethanol since the 1970s. Whereas research and production subsidies were massive until the late 1990s, nowadays sugar-based ethanol has become competitive in price. Also, in recent years concern for the exploitation of sugarcane workers has led to the beginnings of improvements in their working conditions. In parallel, automobile combustion engine technology in Brazil has evolved to mass production and commercialization of flex-fuel engines, which run on any combination of gasoline (fossil fuel) and ethanol (green fuel). Chapter 6 identifies the key advantages and accomplishments in Brazil, and examines the potential for success in other big sugar producers, such as the United States, Colombia, Mexico, and Cuba.

1.8.5 Green Tourism

Tourism is both a large economic activity and an environmentally harmful one. The World Tourism Organization estimates that 5% of total CO_2 emissions have to do with tourism. In small, fragile ecosystems, tourism can improve the livelihood of local residents but also wreak havoc to the environment and damage long-term economic prospects. Tourism practices should be both socially and environmentally sustainable; that is, they should help local communities develop while not harming the environment. Chapter 7 has a regional focus, examining how an important Latin American tourist destination that is extremely sensitive to environmental threats, the Galapagos Islands, is incorporating practices and standards for tourism activities that are beneficial to both the community and the environment. The chapter provides a unique view of the main issues from the point of view of directly affected stakeholders building on primary research conducted *in loco*. It investigates the area's ecosystem; the area's socioeconomic environment, including additional economic activities; the tourism infrastructure, including companies permitted to operate; the role of governmental institutions and regulators; and the impact of visitors in a holistic approach to better understand the influence of these factors and to identify best practices that may be transferred to other environments.

2

The Need for Sustainable Energy Systems

José Gómez Moreno

CONTENTS

2.1 Introduction...18
2.2 Rationale for Sustainable Change...18
    2.2.1 Energy Demand: Absolute and per Capita
        Consumption Will Further Increase.......................................19
    2.2.2 Energy Supply May Not Match Future Demand Growth.... 20
    2.2.3 Oil Addiction: Importers' Economic Performance
        Highly Depends on Energy Prices..22
    2.2.4 Future Energy Cost: Oil Prices Will Become Higher
        and Unstable 23
    2.2.5 Energy Security: Current Policies May Not Secure
        Long-Term Supply..24
2.3 Assessment of Attitudes and Behaviors about Sustainability...........25
    2.3.1 Climate Change...25
    2.3.2 Governments' Sustainability Policies.....................................26
    2.3.3 Consumers 28
    2.3.4 Corporations... 28
2.4 Proposed Guidelines for the Transition to a Sustainable
    Energy System 29
    2.4.1 Energy Efficiency.. 30
    2.4.2 Electricity Infrastructure: Flexibility and Storage 30
    2.4.3 Clean Energy Sources: Nuclear and Renewable 31
2.5 Conclusion 32

2.1 INTRODUCTION

In recent years, discussions regarding the need and means to implement sustainable policies have become widespread. Despite the widely different points of view, stakeholders around the world can roughly be divided into two camps. The first camp considers the "green" movement not only justified but also absolutely necessary, and it demands a drastic change in the management of the planet's resources. Proponents of this point of view stress that governments, consumers, and companies should be aware of the dangers of sitting idly on the sidelines and advocate that governments should establish policies to accelerate the transition to a more sustainable society. The other camp believes that the current sustainability concerns are a temporary phenomenon that will subside as similar past environmental crazes did, and it demands further scientific proof of the influence of human activities on Earth's resources and climate. This view considers that technological development will resolve the resource constraints and the environmental effects of human activities and that the government role in sustainability efforts should be limited, leaving necessary corrections to market forces.

The objectives of this chapter are to outline the facts that support the thesis of sustainability advocates, to examine the attitudes and current actions of the different stakeholders in society regarding sustainability, and finally to propose some guidelines for prioritizing sustainable policies and technologies. The overall emphasis will be placed on the energy sector.

2.2 RATIONALE FOR SUSTAINABLE CHANGE

Some consensus has emerged around the idea that systemic change to make the energy sector more sustainable is required to prevent severe economic and social consequences in the next decades. Five observations from leading research organizations such as the International Energy Agency (IEA), the U.S. Department of Energy, and IHS CERA provide support for such a thesis: first, absolute and per capita energy demand is expected to continue to increase; second, energy supply, both in amount and number of sources, might not increase fast enough to match this increased demand;

third, economic performance of the major economies (most of them oil importers) depends significantly on energy prices; fourth, the current outlook points to a medium-term scenario of increasing energy prices; and fifth, current energy security policies may not ensure long-term access to energy resources. The next five subsections of this chapter sequentially explore each of the above arguments.

2.2.1 Energy Demand: Absolute and per Capita Consumption Will Further Increase

Energy demand is expected to increase significantly over the next 20 years as a result of the growth of the population, the economies, and living standards in developing countries. Projections from the IEA show that global energy demand will increase by almost 40% by 2030, a cumulative annual growth rate of 1.5% (IEA 2009, 74). Several demographic and economic factors account for that significant increase in energy demand. First, according to the United Nations, the planet's population will grow by 20.3% by 2030 (Population Division 2008). The second contributing factor has to do with changes in the population pyramid of developing countries. High birth rates until 10 years ago will increase the working age population (with a corresponding increase in GDP), and with it energy consumption per capita (J. Cohen 2001). The third contributing factor is the urbanization process in developing countries. In recent decades, developing countries have experienced a gradual process of urbanization (during the period 1950 to 2000 developing country urban population increased from 18% to 40%). This urbanization trend is expected to continue over the next few decades, likely leading to an urban population percentage of 60% by 2030 (J. Cohen 2001). Since urban lifestyles require significantly more energy than rural lifestyles because of the higher level of services and the need for transportation, this urbanization process will also impact overall energy consumption.

Finally, economic growth (especially the increase of GDP per capita) will play a key role in energy demand increase. The reference long-term growth rate for world GDP (adjusted for the recent economic downturn) obtained as an average of three rates provided by recognized studies is 3.1% (IEA 2009, 62; OECD 2008, 82; U.S. Department of Energy 2009, 14). Developing countries (especially in Asia) are expected to drive this growth whereas developed economies (with a collective growth rate below 2%) will gradually reduce their aggregate contribution to global GDP.

Analyzing 2030 published estimates of global GDP and population, it is straightforward to ascertain that as a result of these changes, global GDP per capita is expected to grow in the next 20 years by almost 50%. That increase will not be equally distributed: India and China rather than developed countries will account for most of it (175% and 191% respectively, versus 29.5% growth in developed countries) (IEA 2009, Figure 3). As shown in Figure 2.1, there is an approximately log-linear positive correlation between a country's GDP per capita and its energy consumption per capita so every increase in the GDP per capita of a country should be accompanied by an increase in its energy consumption. Therefore, the narrowing of the gap in living standards between developed and developing countries should significantly influence the pattern of energy consumption of the world.

2.2.2 Energy Supply May Not Match Future Demand Growth

Current levels of energy supply, in terms of both quantity and mix, cannot be significantly modified in the medium term. The amount of new fossil resources (especially oil and gas) expected to be added might not be enough to match the growth in demand, given the slowdown of upstream investments, the decline of new discoveries of giant oilfields, and the restructuring of the oil services sector (Jackson 2009, 8, 10–11). Nuclear energy might be partially restricted by the opposition of environmentalist groups or international non-proliferation agreements. Only renewable energy and coal have the potential to grow. The contribution of wind and solar energy will increase significantly, but given that these sources of energy currently represent less than 1% of the primary energy mix, they might not be able to absorb the 40% expected growth of demand. Coal is plentiful in India, China, the United States, and Russia, but it creates severe pollution and carbon emissions, so its use might be limited as well. Therefore, unless there is a significant change in energy consumption or production patterns, or an unexpected technological leap happens, energy supply might not match the expected demand increase and the weight of fossil fuels in the primary energy supply might not be reduced in the medium term. Table 2.1 summarizes IEA's projection of expected shares of the different sources of energy in 2030, considering current expected policies, and compares them to the current shares.

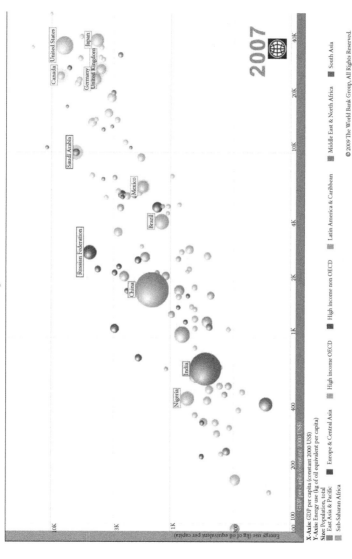

FIGURE 2.1

Log GDP per capita versus log energy use per capita. (World Bank, *Data Visualizer—World Development Indicators 2009*, http://devdata.world-bank.org/DataVisualizer, accessed April 2010.)

TABLE 2.1

Absolute Demand and Share of Primary Energy Mix of the Different Energy Sources, 2007–2030

	Demand (Mtoe)			Share (%)	
	2007	2030	Growth	2007	2030
TOTAL FOSSIL	9,789	13,457	1.4%	81.5%	80.2%
Coal	3,184	4,887	1.9%	26.5%	29.1%
Oil	4,093	5,009	0.9%	34.1%	29.8%
Gas	2,512	3,561	1.5%	20.9%	21.2%
TOTAL RENEWABLE	1,515	2,376	2.0%	12.6%	14.2%
Hydro	265	402	1.8%	2.2%	2.4%
Biomass and Waste	1,176	1,604	1.4%	9.8%	9.6%
Other Renewable (Sun, Wind, Geo)	74	370	7.2%	0.6%	2.2%
NUCLEAR	709	956	1.3%	5.9%	5.7%
TOTAL	12013	16789	1.5%	100.0%	100.0%

Source: International Energy Agency, *World Energy Outlook 2009* (OECD/IEA, 2009), Table 1.1.

2.2.3 Oil Addiction: Importers' Economic Performance Highly Depends on Energy Prices

The importance of energy prices (specifically oil) to the world economy cannot be overemphasized. The price of oil is a reference for the cost structure of virtually all industrial sectors, because oil is massively used for transportation (with no foreseeable substitutes), it determines the prices of other energy sources (such as gas) or the feasibility of energy projects (such as nuclear or renewable), and is a basic input in key industries such as chemicals, plastics, and fertilizers. In the past, unexpected oil price fluctuations have had a deep impact on the major economic variables of oil-importing economies, as in the oil crises of the 1970s and 1980s. Rises in oil price usually precede an increase in inflation, a worsening of an oil-importing country's trade balance, and a slowdown of GDP growth (Birol 2004, 4). Figure 2.2 graphically illustrates these effects for the case of the United States in the period 1970 to 2006.

Nevertheless, the policies of oil-importing countries have not achieved a significant reduction in oil dependency, making their economic performance still highly dependent on oil prices. For these oil-importing countries, ensuring oil supply and price stability has become a priority strategic concern and a significant challenge.

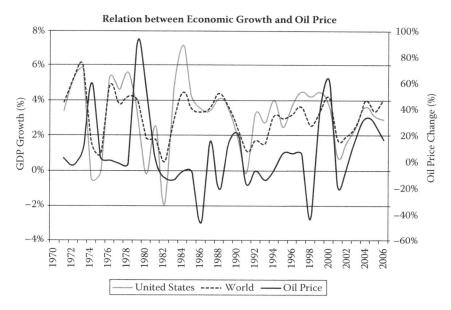

FIGURE 2.2

Oil prices and real GDP growth in the United States and the world. (Elaborated from GDP data from World Bank data and fuel prices from Energy Information Administration, http://web.worldbank.org/WEBSITE/EXTERNAL/DATASTATISTICS and http://tonto.eia.doe.gov/dnav/pet/pet_pri_spt_s1_d.htm.)

2.2.4 Future Energy Cost: Oil Prices Will Become Higher and Unstable

Several arguments support a scenario of increasing oil prices in the medium and long term, according to the IEA. First, world oil demand is expected to rise significantly and importers are expected to maintain or increase their dependency, as Table 2.2 depicts. Second, oil supply is expected to be more concentrated in the OPEC countries, from 44% to 52%, and the Middle East will remain the main source of oil exports (from 49% to 52%) (IEA 2009, 115–116). Third, investment in the oil sector has recently shrunk, as a result of the drop in oil prices caused by the 2008 economic downturn. This reduction in investment may decrease the number of new projects starting operations in the medium term, reducing future excess capacity and thus increasing expected oil prices when demand recovers (Jackson 2009, 10–11). Last, although there are proven conventional oil reserves good for several decades at the current rate of consumption, the average production decline rate in existing oil fields is increasing. As new, more expensive non-conventional projects such as deep sea, extra-heavy oil, and

TABLE 2.2

Dependency on Foreign Oil Imports of Major
Economies in 2009 and 2030

	2009	2030
India	74%	92%
China	50%	74%
EU	81%	91%
Japan	91%	88%
ASEAN	24%	73%
US	63%	58%

Source: International Energy Agency, *World Energy Outlook 2009* (OECD/IEA, 2009), 117.

oil sands come on-stream to meet growing demand, average oil extraction costs will increase (IEA 2008b, 218).

2.2.5 Energy Security: Current Policies May Not Secure Long-Term Supply

Given the uncertainties surrounding oil supply and price, in the past four decades major oil-importing economies have established energy policies that attempt to ensure oil supply. The analysis of the energy security policies of major oil importers (OECD countries, China, and India) reveals that future oil supply may not be guaranteed. First, oil importers cannot further diversify their supply, as it is becoming increasingly concentrated. Another barrier to energy security is the lack of international collaboration. Consumers (United States, European Union, China, and India) and producers (Middle East, Venezuela, Russia) show limited cooperation in any international affair (including energy policy); this limited cooperation is due to differences in strategic objectives and ideological makeup. In addition, current international emergency oil security policies exclude China and India, hindering the possibility of coordinated action in case of disruption. Last, major oil-exporting regions (Middle East, South America, and Africa) suffer from endemic political instability that shows few signs of improvement (IEA 2009, 117–118; Yergin 2009, 8–10).

The conclusion of this analysis is that unless major oil-importing countries and producers take specific multilateral action for promoting collaboration and coordinated strategies, there might be inherent oil supply instability with the possibility of oil price surges.

2.3 ASSESSMENT OF ATTITUDES AND BEHAVIORS ABOUT SUSTAINABILITY

To assess the current momentum of the sustainability movement, it is necessary to examine the attitudes and behaviors of governments, consumers, and corporations. Climate change debates in the past decade have introduced the sustainability factor into the policy-making process of governments and gained the attention of consumers and corporations. But according to research from Havas, Mintel, and McKinsey, this awareness is yet to be translated into specific policies and actions that will alter the management of resources and the behaviors of consumers and companies in the short and medium term.

2.3.1 Climate Change

The possibility of structural and lasting climate change caused by human activity is by now widely accepted in government and media circles and has turned into a significant driver for the sustainability movement. Although there is some debate among scientists on different issues such as the scientific evidence of climate change or the role of human activity in it, the United Nations and the majority of developed countries have accepted its existence. Some governments are introducing climate change–related policies (both at a national and an international level) based on its potential catastrophic consequences. *The Stern Review*, published in 2007 by a team of economists at the British government, quantified the long-term effects of climate change on welfare at a 5% to 20% reduction of GDP per capita (see Figure 2.3). Although those figures are subject to forecasting error, the likely order of magnitude of the consequences is significant enough for governments to take action to address it (Stern 2007).

It is widely believed that in order to avoid catastrophic consequences, the planet's average temperature should not increase by more than 2° to 3°C. In the IEA's "business as usual" scenario, global carbon emissions are expected to grow by 39.5% from 2007 to 2030, mostly in developing countries in Asia (developed economies are expected to reduce their carbon emission levels through 2030) (IEA 2009, 623, 625, 647, 649, 651). The Intergovernmental Panel on Climate Change (IPCC) report estimates that such emissions would raise the carbon dioxide equivalent in the atmosphere to a level between 710 and 855 parts per million (ppm) (currently

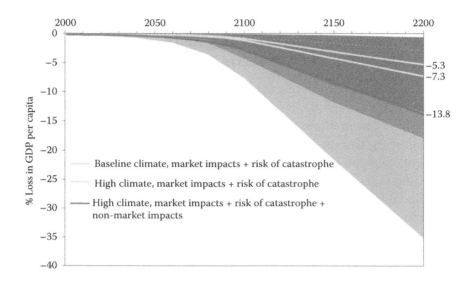

FIGURE 2.3
Three scenarios of British GDP loss due to climate change. (Stern, Nicholas, *The Economics of Climate Change: The Stern Review*. Cambridge: HM Treasury, 2007, 157.)

it is approximately 430ppm), raising the planet average temperature by up to 6°C, significantly higher than the 2°C safety threshold. Such limited warming can only be reasonably achieved by stabilizing CO_2 levels in the atmosphere between 450 and 550 ppm, which implies a significant reduction of emissions (30%–85%) at a cost of at least 1% of the world's GDP by 2050 (IPCC 2007, 15; Stern 2007).

2.3.2 Governments' Sustainability Policies

Every major economy is starting to incorporate sustainability into their policy-making agendas, although at different paces and toward different specific goals depending on their development level and the impact on the economy. Developed countries are focused on establishing carbon trading mechanisms and reduction objectives, developing new technologies such as carbon sequestration or the smart grid, and installing renewable capacity. Developing countries are more focused on solutions that would allow them to maintain their growth (such as increasing energy efficiency) and on creating their own sustainable industries in order to compete in global markets. Table 2.3 shows the sustainability-related policies that the United States, the European Union, and China are putting in place.

TABLE 2.3

Major Climate Change Policies in the United States, EU, and China

	United States	**EU**	**China**
Programs	Waxman-Markey Bill	EU Energy Security and Solidarity Action Plan, 2nd Review	Chinese Five Year Plan 2006–2010
Objectives	Efficiency, global warming, clean economy	Sustainability, competitiveness, and security of supply	Sustainable development
Emission Reductions	17% by 2020	20% by 2020	Non-binding 40-45% reduction of carbon intensity by 2020
Carbon Trade	Yes, to be established	Yes, in operation	No
Renewable Share	20% of electricity by 2020	20% by 2020	15% by 2020
Energy Efficiency	Improve energy productivity by 2.5% per year until 2030	Improve 20% by 2020	Improve 20% by 2010

Sources: European Commission, EU Energy Security and Solidarity Action Plan: 2nd Strategic Energy Review Summary (2008), http://www.europa-eu-un.org/articles/en/article_8300_en.htm; House of Representatives, HR2454, American Clean Energy and Security Act of 2009 (June 26, 2009), http://thomas.loc.gov/cgi-bin/bdquery/z?d111:HR02454:@@@D&summ2=m&; Zhang, ZhongXiang, *Assessing China's Energy Conservation and Carbon Intensity: How Will the Future Differ from the Past?* (May 24, 2010), Working Paper, East-West Center, Honolulu, HI.

While these countries are taking measures to increase sustainability at the national level, international agreements are evolving at a much slower pace. The last climate change meeting, held in Copenhagen in December 2009, did not result in any significant agreement when compared to the expectations created prior to the summit. Many countries signed a nonbinding agreement geared toward limiting temperature increases to under 2°C, but no party accepted any binding commitment on emissions reduction or any specific control mechanism. One outcome of the Copenhagen meeting was the perception that any further international binding agreement will likely require a direct understanding between the United States, China, and India (the biggest emitters in the near future) since United Nations–led multilateral efforts had to date been unsuccessful.

2.3.3 Consumers

Another significant driver of the sustainability movement is consumer awareness of sustainability and climate change issues. According to a 2008 IPSOS consumer survey for Havas Media, 78% of consumers believe in an anthropogenic climate change and 81% agree that a change in lifestyle will be necessary to combat it. The survey found that 89% of consumers expected to purchase more sustainable products in the next year and that 79% of consumers would endorse companies perceived as environmentally and socially responsible (Havas Media 2008, 3–4).

This awareness and positive attitude of consumers toward sustainable products contrasts with their actual behavior as revealed by Mintel's 2009 *Green Living* report. According to this report, although sustainable product demand is increasing, only 11% to 30% of U.S. consumers considered sustainability factors in their last major purchase (e.g., house, car, appliances, etc.). In the service sector, green criteria are considered by only 10% to 16% of users (Mintel 2009). Two reasons may account for this divergence between awareness and behavior. First, although sustainability might be an appreciated attribute in isolation, it could be secondary when compared to other attributes such as price or quality. The small premiums (15% or less, according to the Mintel report) that consumers are willing to pay for sustainable products may confirm this argument. This behavior may have become more marked since the recent economic downturn, as consumer purchasing power shrank. Second, the reduced availability of sustainable options in some product categories and the lack of adequate information and labeling may be limiting the ability of consumers to purchase sustainable products (Mintel 2009).

In order to make sustainable products widespread in the future, these two barriers should be overcome. Sustainable products would need to match the price and quality of their non-sustainable alternatives, and businesses would have to provide consumers with adequate offerings and specific information about sustainable products.

2.3.4 Corporations

Companies are also realizing the relevance of sustainability, according to a McKinsey Quarterly Survey of 2,192 executives. Sixty percent of the survey respondents found that climate change is an important element to consider in overall corporate strategy, and 61% considered that good management of climate change issues would result in positive effects on companies' bottom lines. More than 50% of the respondents said climate change was an important element to consider in new product development, investment planning, management of environmental issues, and procurement (McKinsey 2007, 3–8).

Similarly to consumers, there is a gap between the intentions of corporations and their actual actions. The McKinsey study pointed out that about 40% of those companies never took into account climate change in their operations and that 44% of the CEOs had said that climate change was not a significant item in their agendas. Regarding the factors favoring consideration of climate change by companies, corporate reputation and media attention seem to have been more important than consumer preferences or personal convictions (McKinsey 2007, 3–8).

This divergence may reveal that the incentive system within which corporations currently operate might not be oriented toward sustainability. Although consumers prefer sustainable products and are willing to favor sustainable companies, these cannot directly benefit because customers reject price premiums (necessary to compensate for higher costs) and because regulation does not reward sustainable behavior (or punish the alternative). The debate is whether the government should step in and if so, which specific mechanisms should be used to align the corporate incentive system with sustainability (subsidies, markets, taxation, environmental regulations, labeling, energy standards, etc.). Governments of countries that have achieved greater sustainability improvements usually took a significant role in shaping incentive systems. Given that these countries are mostly European nations with welfare systems, it remains to be seen how the model might transfer to different political and socioeconomic systems such as those of the United States and Asian countries. Chapter 3 presents the case of Denmark, describing how that country effected lasting change in the incentive systems of both consumers and companies to make energy generation, distribution, and use more sustainable.

2.4 PROPOSED GUIDELINES FOR THE TRANSITION TO A SUSTAINABLE ENERGY SYSTEM

As described in previous sections, the world is facing significant sustainability challenges. The high complexity of those issues and the wide scope of policies and technologies available to governments, companies, and individuals require a general framework to prioritize resource allocation. The following guidelines attempt to provide structure to the identification, analysis, and prioritization of the measures that would enhance overall energy system sustainability.

2.4.1 Energy Efficiency

Ever-increasing energy efficiency and conservation should be a priority for governments, companies, and consumers. Energy efficiency and conservation policies can have a very positive effect on the economy, increasing the competitiveness of companies (through reductions in energy costs), improving the trade balance disequilibrium by reducing fuel imports, avoiding or delaying capital investments in infrastructure, and creating employment (Janssen 2008, 4). Increasing energy efficiency also exhibits more attractive economic characteristics than alternative sustainability-enhancing actions because the reduction in energy costs allows for faster investment recovery and because it is more cost effective to reduce inefficiencies in energy production and consumption than to expand the capacity of existing energy infrastructure, whether conventional or renewable.

On the supply side, energy efficiency measures affect power generation, transportation and distribution, district heating, and public transportation. The degree of implementation of these supply-side measures depends on the specific electric model of the country and the incentive system of utilities and producers. On the demand side, energy efficiency measures focus on industrial and residential users, buildings, and vehicles. Business users tend to be more proactive in implementing energy efficiency and conservation measures, especially when energy is a significant part of their cost structure or there are environmental requirements. At the same time, individuals seem more reluctant to adopt energy conservation because of the relatively small weight of electricity bills in household budgets, the need for upfront investments, and the challenge of modifying behaviors usually embedded in their respective sociocultural settings.

2.4.2 Electricity Infrastructure: Flexibility and Storage

Although energy efficiency can provide significant improvements toward sustainability, it is also necessary to incorporate technologies that can provide cleaner means to generate and use energy, such as renewable sourcing technologies (such as wind and solar, supplied by both utility-scale farms and decentralized small-scale individual producers), real-time management of home appliances through demand-response applications, electricity storage at the utility and user levels, and hydrogen fuel cells for home consumption and transportation, to cite a few examples. These sustainable technologies are small scale, intermittent (onshore wind and solar energies depend on meteorological conditions and therefore are intrinsically variable), and able to supply or absorb power depending on the circumstances. These characteristics require that the electricity infrastructure be flexible enough to control myriad producers and consumers at any given time without putting the whole system at risk. The current electricity infrastructure was designed decades ago, resulting in a centralized system in which the energy relationship was unidirectional (from producer to consumer), so it is not prepared for the new needs and requirements that characterize clean technologies (Risø National Laboratory 2005, 4; Risø National Laboratory 2009, 5). Some utilities in the United States and Europe have started developing these networks, such as the EDISON project conducted by DONG in Denmark.

A key part in upgrading the energy grid is storage. Electricity demand is unevenly distributed during the 24-hour period (peaking during the daytime) and requires significant extra installed capacity to serve only the peak period, thus elevating the total cost of electricity. In contrast, wind production usually peaks at night and must be stopped often because the grid does not require much energy during the non-peak period (in countries such as Denmark, this problem is limiting the potential contribution of renewable sources to the electricity mix). Introducing storage in the power infrastructure would make stored energy available at the times when it is most needed, reducing the need for extra installed capacity in the system and thus decreasing overall energy cost. Although the price of electricity storage is still very high, new technologies hold promise for the development of cost-effective storage systems, whether utility-sized or distributed small-scale ones (Mason 2008; Risø National Laboratory 2009, 5).

2.4.3 Clean Energy Sources: Nuclear and Renewable

With optimized energy generation and consumption and a flexible power infrastructure in place, the next step should be to focus resources on replacing existing fossil fuel generation capacity with alternatives that minimize environmental impact, decrease energy dependency, and can be scaled up, namely nuclear and renewable energy sources.

Renewable energy sources (such as wind, solar, geothermal, or waves) can provide a very effective and clean solution to take advantage of existing local resources, and replace fossil generation. The European Renewable Energy Council (2004, 2) estimates that renewable energy could provide 50% of the world's primary energy by 2040. Although a 50% contribution may look too high in the medium term (in 2008 solar, wind, and geothermal together contributed less than 1%), in some countries such as Denmark renewable sources already account for almost 20% of electricity production (Danish Energy Agency 2008, 3). The major barriers to the widespread incorporation of renewable energy sources are the lack of cheap electricity storage to address the variability and intermittency of renewable sources, the need to make the electricity system more flexible and modular, and the higher cost of renewable energy (although the cost of onshore wind power and biomass are reasonably close to grid parity, the cost of solar or wave energy is still very high, especially compared to nuclear or coal). Governments, businesses, and investors are pouring a significant amount of resources into the R&D and installation of clean energy sources, so the importance of these technologies will grow in upcoming years, as they become an integral part of the overall energy system.

Nuclear energy can complement some of the shortcomings of renewable energy sources. Nuclear is the only technology that can provide emissions-free, base-load power at prices that are competitive with the fossil fuel alternative (coal). Currently 54 nuclear plants are under construction in the world, 40 of which are in developing economies (IEA 2009, 160). OECD countries are also reviving their nuclear programs. For example, the U.S. administration has granted loan guarantees for building two nuclear plants in Georgia, the British government has established new nuclear sites to build new plants, and both Italy and Sweden have overturned their nuclear bans. Although public opposition could delay some of these new projects, the nuclear revival is expected to gain momentum in coming years.

2.5 CONCLUSION

There are solid arguments to justify changing the current energy system in order to ensure its long-term sustainability. The mismatch between demand and supply, the importance of oil prices to the economy, the medium-term scenario of expected increasing prices, and the uncertainty about medium-term availability of energy resources might put at risk the significant economic, political, and social development achieved in recent decades. A sustainable change in the energy system is thus required for continued economic growth.

There are signs that governments, corporations, and consumers are realizing the need for a sustainable change, based on either energy security or climate change reasons. The major economies are establishing national policies to deal with climate change and to ensure the supply of energy at stable prices. At the same time, consumers and companies are increasingly aware of sustainability and climate change issues and are recognizing the need for changes in lifestyle, although there is a gap between their declared intentions and their actual behavior. The specific measures that governments will implement to change the current incentive system remain one of the main challenges. Among the many options available to policymakers to increase the sustainability of the energy system, three of them should be prioritized: the spread of energy efficiency to the generation, distribution, and use of energy; the upgrade of the electricity grid; and the incorporation of nuclear and renewable energy sources to the system.

3

The Transition from Oil Dependency to Sustainability in Denmark

José Gómez Moreno

CONTENTS

3.1 Overview: Why Is Denmark a Success Story? 36
    3.1.1 Decoupling of Economic Growth from Energy Consumption 37
    3.1.2 Reduction of Energy Needs, Foreign Dependency, and CO_2 Emissions 37
    3.1.3 Development of a New Clean Technology Industry 38
3.2 Key Elements of the Danish Energy System 38
    3.2.1 Energy Policy Continuity...38
    3.2.2 Energy Efficiency and Conservation 41
    3.2.3 The Development of the Network of Combined Heat and Power (CHP) Plants... 42
    3.2.4 The Development of the Wind Sector 43
        3.2.4.1 Grassroots Support and Communities 44
        3.2.4.2 Government Coordination and Incentives 45
        3.2.4.3 Collaboration among All Industry Actors 45
        3.2.4.4 Incremental Development of Technology 46
3.3 Conclusion 46

Energy scarcity and higher energy resource prices, together with enhanced awareness of environmental sustainability issues, are prompting governments, consumers, companies, and international organizations to question certain tenets of current energy and economic models. Several countries have started to alter their environmental and economic policies to ensure sustainability, although not all countries are on the same stage of evolution on the sustainability ladder. Though every country has its own natural endowments, climate, history, culture, economy, political system, and institutions, there is no question that some important lessons are transferable and that there are many opportunities to learn from nations that have made progress along the road to sustainability. Denmark, a small country in Northern Europe with just 5.5 million inhabitants, is an example of such a nation (Central Intelligence Agency 2010).

Because Denmark started early to adopt sustainability policies and is already showing encouraging results, it can provide valuable guidance regarding the transition from a fossil fuel-dependent society to an environmentally sustainable one. Furthermore, Denmark's example can serve as inspiration to countries hoping to use this transition to stimulate economic progress. This chapter analyzes the most relevant steps in the development of Denmark's energy system and examines the unique socioeconomic and political factors that made it possible for the country to become a sustainability role model for the world.

3.1 OVERVIEW: WHY IS DENMARK A SUCCESS STORY?

The two oil crises of the 1970s had a significant impact on the economies of oil-importing countries. At the time, Denmark depended almost entirely on foreign oil supplies for its energy needs. These crises prompted the introduction of the first sustainable policies in Europe, Japan, and the United States, which included diversification of fuel supply, energy conservation, and alternative energy R&D incentives. Whereas some countries partially abandoned these policies once oil prices returned to reasonable levels, some other countries (admittedly a minority) maintained or expanded those policies to seize the opportunity to reduce their energy dependency. Denmark was one such country, and it can be considered a success story because in recent decades it has achieved three noteworthy results: decoupling its economic growth from energy consumption; reducing its energy consumption, carbon emissions, and foreign energy dependency; and developing a thriving clean technology sector.

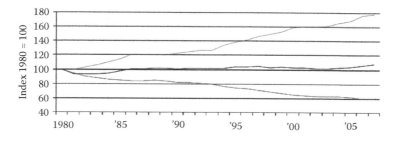

FIGURE 3.1

Denmark energy consumption (top), GDP (middle), and energy intensity (bottom). (Ministry of Climate and Energy of Denmark, *"The Danish Example": The Way to an Energy Efficient and Energy Friendly Economy*, 2009, Copenhagen: Ministry of Climate and Energy. With permission.)

3.1.1 Decoupling of Economic Growth from Energy Consumption

In the period 1980 to 2009, the Danish economy performed well, with its GDP increasing by 78% (2.9% CAGR), while its overall energy consumption remained approximately stable (Ministry of Climate and Energy of Denmark 2009). In the same period, aggregate GDP growth of OECD countries was 112%, while energy demand increased by 46.9% (IEA 2008, 66; OECD 2010). Figure 3.1 graphically depicts the decoupling of Denmark's GDP growth from its energy consumption.

3.1.2 Reduction of Energy Needs, Foreign Dependency, and CO_2 Emissions

Since the 1970s, Danish energy efficiency and carbon emission parameters have improved remarkably. Energy intensity (defined as the energy necessary to produce one dollar of GDP) has been reduced by 42.6% in the period 1980 to 2008. Carbon emissions intensity has been reduced by 34.9% and carbon emissions per capita decreased by 12.2% between 1990 and 2008 (World Energy Council 2008). Foreign-energy dependency has also been solved. Denmark depended almost completely on foreign oil imports (93% of total consumption) for its energy needs in 1973 (Hammar 1999, 3). Currently, after 35 years of sustained policies, Denmark is the only net exporter of energy in the EU (Eurostat, European Commission 2008).

3.1.3 Development of a New Clean Technology Industry

Denmark's energy policies not only reduced emissions and dependency on foreign energy sources and emissions but also provided a significant stimulus to the Danish economy. According to the Confederation of Danish Industry, between 1997 and 2008 Danish exports of clean technologies grew by 265% to 8.6 billion Euros, representing 11% of total commodity exports; the growth rate of clean-tech exports was three times higher than the growth rate of total Danish exports and twice the growth rate of European Union exports over the same period (Confederation of Danish Industry 2010, 10–11).

Employment in the clean technology sector stood at 29,000 full-time employees in 2009 and was growing faster than the country average (16% from 2004 to 2006, while across industries employment grew by 3%). Regarding innovation, the production of sustainability research articles is one of the highest in the world (70 per million inhabitants compared to 16 and 17 in the United States and European Union, respectively) (Confederation of Danish Industry 2010, 11).

3.2 KEY ELEMENTS OF THE DANISH ENERGY SYSTEM

The main goals of Denmark's energy policy are to achieve security in energy supply and to continuously reduce environmental impact. Four factors seem to have been crucial for its success: first, the gradual and coherent roll-out (and continuity) of policies that have evolved from a sole focus on security of supply to include environmental protection; second, the implementation of energy efficiency and conservation policies; third, the development of a network of decentralized small-scale electricity and heat producers; and finally, the promotion of renewable sources of energy (Ministry of Foreign Affairs of Denmark 2008, 3–4). We shall analyze each of them in turn.

3.2.1 Energy Policy Continuity

The continuity and coherence of the successive Danish energy plans over time is arguably the most significant factor in the sustainable transformation of Denmark. Danish government formulated new policies on the basis of the objectives and strategies set by prior energy plans, adapting them to new circumstances, such as increased evidence of climate change or technological advances. This continuity was key to reducing investor and industry uncertainty, to implementing programs that showed results only in the long run (such as fuel substitution, heat and electricity networks, or building codes), and to enabling behavioral changes in households and companies. Table 3.1 outlines the main policies and pieces of legislation during the period from 1976 to 1999 (OECD 1999).

Denmark's modern energy policies started after the first oil crisis of 1973. The embargo severely affected Western countries, forcing Denmark to engage in a difficult adaptation to the new situation, which required visible changes in lifestyles, including bans on Sunday car use, commercial lighting restrictions, and energy saving campaigns. These emergency measures were lifted gradually as oil prices came down, but the Danish government realized that a comprehensive set of policies was necessary to avoid similar consequences in the event of a new oil crisis (IEA 2008a, 2). The government created the Denmark Energy Agency, which coordinated the energy policy efforts of different ministries and issued the 1976 Danish Energy Policy. This policy focused mostly on security of supply to reduce dependence on foreign oil by replacing oil-based electricity generation with other fuels (in less than 10 years, coal represented 95% of electricity generation fuel sources), introducing nuclear power, opening the development and exploitation of Denmark's North Sea oil and gas resources, and allocating funds to alternative energy (Lund 1999, 118; Meyer 2007, 349; OECD 1999; Pedersen 2005, 41–42).

This 1976 policy was followed by the Heat Supply Act of 1979, which introduced a bottom-up planning for the development of an efficient heat supply system (IEA 2010, 7). The surge in oil prices during the 1979 oil crisis influenced the 1981 Energy Plan, which was focused on lowering the cost of energy (maintaining the energy security objective) in order to reduce the energy bill of end users and to improve macroeconomic indicators such as the balance of payments (Pedersen 2005). The 1990 "Energy 2000" plan added environmental sustainability (specifically taking climate change into account) as an important policy focus (Lund 1999, 117–118). The major objectives of this plan were the establishment of targets for the reduction of CO_2, SO_2, and NO_2 emissions by 2005, the transition to cleaner fuels (expanding the use of natural gas, biomass, solar, and wind), and the increase in energy efficiency through decentralized combined heat and power plants, improvements in building codes, labeling, and energy

TABLE 3.1

Denmark Energy Plans and Related Legislative Acts, 1976–1999

Year	Act or Plan	Main Purpose or Objective
1976	Electricity Supply Act	Governs the development and structure of the electricity sector
1976	Danish Energy Policy 1976	Security of supply, energy savings, oil substitution
1979	Natural Gas Supply Act (called elsewhere Construction of the Natural Gas Project Act)	Governs the development of the natural gas network
1979	Heat Supply Act	Governs the development of the district heat sector
1981	Energy Plan 1981	Security at lowest cost through substitution of imported fuels: convert large power plants from oil to gas, more CHP, gas, and renewables
1990	Energy 2000	CO_2, SO_2, and NO_2 reduction targets for 2005, promotion of renewable energy and CHP, global environmental initiatives, energy savings
1990	Heat Supply Act (Amended)	Framework for converting district heat to CHP, and heat pricing (Danish Energy Agency 1998)
1992	Carbon Tax legislation	Detailed tax scheme, investment subsidies to energy efficiency measures, subsidies to CHP and renewable energies
1996	Energy 21	Targets for CO_2 reductions for 2005 and 2030, further promotion of CHP and renewable energy
1996	Electricity Supply Act (Amended)	Came into force January 1, 1998 Governs structure and economic regulation of the electricity sector "with a particular aim to promote the environmentally benign utilization of energy"
1999	Energy Supply Act	Adopted on May 24, 1999 Introduces competition into production and trade while maintaining the objectives of the 1996 Electricity Supply Act

Source: OECD, Denmark *Regulatory Reform in Electricity* (1999), Table 1.

standards for appliances (OECD 1999, 9; Pedersen 2005). Energy 2000 set the basic guidelines of the ensuing Danish Energy plans—"Energy 21" (1996), "Climate Strategy" (2003), and "A Visionary Danish Energy Policy" (2007)—which have expanded the objectives of Energy 2000 to achieve ever more ambitious emission reductions and to increase the contribution of renewable sources.

3.2.2 Energy Efficiency and Conservation

Since the 1970s Denmark has achieved one of the most significant reductions in energy and carbon intensity in the world, decreasing its energy costs, environmental impact, and foreign energy dependency. The residential sector accomplished the most significant improvements, reducing its energy and carbon intensity by 45.1% and 43.2% (1990–2005 period), respectively. Improvements in efficiency of manufacturing processes led to a 30% decrease in the industrial sector's energy intensity and a 14.2% decrease of its carbon intensity. The transportation sector also improved its energy and carbon intensity by 10.2% and 8.4%, respectively (World Energy Council 2007).

Denmark introduced the first specific energy efficiency policies after the first oil shock, when the government realized the need to reduce foreign energy dependency. One focus of these measures was on energy conservation in households, which represented 31.2% of final energy consumption in 1990 (Statistics Denmark). These programs comprised energy conservation campaigns through public and private institutions, additional taxes on oil and electricity, a program of grants to stimulate energy conservation building retrofitting, and stringent energy standards in building codes (Olesen n.d., 1; Schipper 1983, 321–322).

Two of the most significant barriers to the expansion of energy efficiency were the lack of standards and the limited product offerings available in the market. The government attempted to increase the supply of energy-efficient houses by introducing labeling programs, which subsidized energy audits in existing buildings and required energy efficiency certificates for real-estate transactions. This measure raised the amount of information in the real estate market, providing Danes with greater availability of green options for renting and buying properties. Labeling was mandated for heaters and appliances as well (Olesen n.d., 1; Schipper 1983, 321–322).

The Danish government also implemented policies intended to reduce the energy consumption and CO_2 emissions of the Danish industrial sector. In 1996, the government introduced a "green tax" package, whereby companies would be taxed on their CO_2 emissions. This package included the possibility for CO_2-intensive industries to reach a binding agreement with the government in order to receive a preferential CO_2 tax rate in exchange for developing and implementing a program of energy efficiency measures (Hansen 2001, 3–4).

The transportation sector accounted for 28.8% of energy consumption in 1990 (Statistics Denmark). Denmark's measures to reduce energy consumption and emissions in the transportation sector have been to institute taxes (both to fuel and to vehicles) and to promote new transportation technologies such as electric vehicles.

3.2.3 The Development of the Network of Combined Heat and Power (CHP) Plants

In 1960 Denmark had only a few large-scale electric plants that provided for the country's electricity needs and whose residual heat was used for centralized district heating in the larger cities. The rest of the country fulfilled its heating needs by using local district heating (mostly owned by municipalities and cooperatives) or individual boilers that consumed heavy oil (International Energy Agency 2008a, 1). The Heat Supply Act of 1979 instituted a planning process for the development of a heat supply system in the country, following a bottom-up approach from the local to the regional level. Local authorities had to determine their heating needs and establish the plans to fulfill those needs. Those plans then were aggregated at the county and regional levels in a structured approach that allowed for the optimization of resource allocation in the spread of district heating (Danish Energy Authority 2005, 15; International Energy Agency 2010, 7).

During the 1980s, the Denmark Energy Agency developed programs to disseminate CHP use. CHPs are generation units that produce both electric power and heat for houses and businesses more efficiently than standard alternatives. CHPs produce higher output per unit of fuel, given that they are designed to recover the thermal energy lost in the electricity generation process and utilize it for heating. The 1986 Agreement on Cogenerated Heat and Electricity promoted the expansion of the decentralized CHP network significantly (Danish Energy Authority 2005, 11, 15). As an incentive for the installation of these systems, the regulatory agency introduced measures such as taxes on heating fuel (except when used for CHPs), CHP electricity feed-in tariffs (a subsidy consisting in the grid's obligation to buy electricity from these plants at a given price), and premiums for CHP electricity generated with biomass and biogas (International Energy Agency 2010, 7–8). The availability of subsidies and the ownership structure (local heat district entities remained as cooperatives or non-profit municipal utilities, which ensured that energy prices would remain fair and that any savings would revert to their members) prompted many small and mid-size heating districts to adopt CHP technology (Hammar 1999, 4; International Energy Agency 2008a, 3; Maegaard 2009, 4). The replacement of existing boilers and electric plants for CHPs increased overall systemic efficiency and enabled the use of less-polluting fuel sources, reducing energy consumption per unit of residential surface by 50% and cutting CO_2 emissions per year between 8 to 11 million tons (Danish Energy Authority 2005, 16; International Energy Agency 2010, 10).

The 1999 and 2004 restructuring of the electricity sector laid the foundations for a decentralized power system and allowed for the incorporation of renewable energy in ensuing decades. The restructuring allowed any private or public producers (including local community-owned CHPs or wind farms) to engage in energy generation and trading. The transmission grid was opened to those producers so they could join the market. Electricity distribution remained in the hands of local not-for-profit cooperatives, municipalities, and concessionaires. Lastly, corporate unbundling was introduced, reducing the maximum stake of generators and retail distributors in system operators and transmission companies to 15% (Maegaard 2009, 9; OECD 1999, 7, 30–31).

All of these efforts resulted in a transition from a system with a few centralized electric plants and heat-only district heating networks in the 1970s to the current decentralized network of CHP plants and windmills. This transition completely changed the landscape of electric and heat generation in Denmark, as Figure 3.2 shows.

3.2.4 The Development of the Wind Sector

The success of the development of the Danish wind power sector, which generates almost 20% of the electricity in Denmark and is a major player in the world's turbine market, is the result of several factors, including two mutually reinforcing movements, one bottom-up (citizen initiatives)

Centralized production in the mid 80's   Decentralized production of today

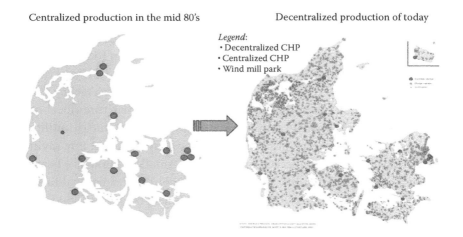

FIGURE 3.2
Denmark electric generation system in the mid-1980s and 2010. (Danish Energy Agency, *Denmark: A Leading Player in Combined Heat and Power*, n.d., http://www.ens.dk/en-US/ Info/news/Factsheet/Documents/kraftvarme%20170709.pdf, accessed April 2010. With permission.)

and another top-down (government policies); the intense collaboration among all of the stakeholders in the industry; and the incremental technology development.

3.2.4.1 Grassroots Support and Communities

The widespread grassroots support of different civil society groups was one key factor for the successful development of wind energy in Denmark. Since the 1970s, environmental groups started gaining public support and began shaping the Danish political agenda. This process culminated in 1985, when the Danish Parliament rejected the nuclear program. Environmental groups also supported alternative energy sources, prompting the government to introduce legislation supportive of renewable energy (Christianson 2005, 1–2; Hvelplund 2006, 3; Meyer 2007, 349).

The existence of cooperative groups and communities owning CHP plants in the 1980s and 1990s also provided important support for the development of wind energy technology, as they installed many of the original wind-generation turbines. Two reasons may justify these groups' early adoption of the new technology. First, wind power has a small or medium scale and is decentralized (wind potential is sparsely distributed), fitting the needs and characteristics of communities and cooperatives

(Christianson 2005, 3). Second, the incentives that the government originally provided to community CHPs were extended to the wind-energy pioneers (Hvelplund 2006, 3).

3.2.4.2 Government Coordination and Incentives

The government's lead in coordinating industry efforts allowed for rapid technology deployment and for rapid movement along the learning curve. Risø DTU, the National Laboratory for Sustainable Energy at the Technical University of Denmark, was transformed into the main test center for the nascent wind industry. The government required new turbine performance testing and certification at the national test station in order for subsidies to be given (Boon 2008, 50; Christensen et al. 2005; Kjems, 2009, 61).

The importance of the incentives provided by the Danish government to wind energy–related initiatives cannot be over-emphasized. Government incentives comprised subsidies for the installation of wind turbines (covering 30% of the investment cost), mandated utility purchases of all generated wind power at a fixed price (85% of market price), and offered tax breaks for wind-generated electricity (Hvelplund 2006, 3). Under this scheme, developers and communities had an incentive to install new turbines, given the possibility of profits (to be spread among community members) and tax breaks. As citizens benefited directly from subsidies, they were quite supportive of wind farms and did not show significant opposition, contributing to the rapid expansion of turbines throughout the country (Maegaard 2009, 16).

The government also had long-term involvement in funding basic R&D for the industry through Risø, which played a very significant role in the development of key wind technologies, such as the identification and quantification of wind resources (Breakthrough Institute 2009, 22).

3.2.4.3 Collaboration among All Industry Actors

Another key factor in the successful development of the Danish wind power industry was the strong collaboration among manufacturers, developers, operators, scientists, and regulatory authorities. For example, research institutions focused on solving short-term problems faced by developers and operators. At the same time, the wind industry funded basic research connected to wind energy, so scientists gained substantial knowledge regarding basic design questions (Kjems 2009, 61). Risø also fostered the exchange of information among different actors by making test results and operational turbine data available. Risø's interest in being fully informed on technology evolution reduced the bureaucratic burden to applied research and development (Boon 2008, 50; Christensen et al. 2005; Kjems 2009, 61). This multilateral collaboration enabled the industry to gain a competitive edge and scale up to more efficient designs.

3.2.4.4 Incremental Development of Technology

Several small Danish equipment manufacturers entered this new industry trying to meet the needs of their individual and cooperative customer bases (Christianson 2005, 4). The first turbine designs were small-sized, customized to each individual application, and did not require unmanageable investments. The turbines grew in size and complexity following an incremental process of technological development that played a crucial role in the success of the nascent Danish wind industry. Other countries at that time relied on bigger corporate players to develop the industry and focused on more complex technological designs, usually requiring larger investments and higher risk-taking, which did not lead to more successful results (Boon 2008, 25; Meyer 2007, 354).

3.3 CONCLUSION

This chapter described the success that Denmark has achieved in adopting an energy sustainability model. The Danish experience could be both scalable to larger and more populous countries, and transferable to other cultures in different parts of the globe. At least five lessons seem important and applicable to other environments.

It is possible to combine economic growth with sustainability. The case of Denmark demonstrates that a country can increase its long-term sustainability and achieve high levels of welfare, economic growth, employment, and innovation.

There is no silver bullet. No single technology by itself can solve the problem of sustainability. Denmark increased energy efficiency, installed CHP systems, and exploited biomass and wind resources. Governments should analyze the optimal combination of technologies given their respective specific political, cultural, geographic, socioeconomic, and technological circumstances, and then carefully plan their implementation schedule in order to optimize resource allocation.

Long-term policies must be maintained over time. Following the oil crisis, unlike many other countries, Denmark succeeded in cutting its oil dependency and in reducing the resulting negative environmental impact. The difference was that Denmark recognized the long-term implications of those policies and put in place mechanisms to sustain them over time. Although Denmark eventually adapted these policies to changing circumstances, there was consistency in the long-term strategic objectives with a well thought-out incentive system that provided the right signals to various economic agents and remained relatively stable over time.

Collaboration among government, investors, researchers, and industry is needed. Although the Danish government played an important role in leading, coordinating, and providing incentives to the wind industry, the significant collaboration between different stakeholders allowed for faster gains along the learning curve and facilitated the rapid expansion of the new technology in the national market. In order to achieve such a degree of collaboration, it was necessary to establish a framework of participation and to set the right incentives for all the parties to collaborate, especially when interests conflicted.

The incentive system should attract the participation and support of citizens. In Denmark there was strong public participation—originating in a tradition of community organization and civic involvement—supporting environmentalism. The government responded to this grassroots demand with an incentive system that rewarded sustainable behavior on the part of the general public and helped create and increase aggregate demand for renewable energy.

4

The Revival of Battery-Powered Electric Vehicles in Japan

Jan Zelezny

CONTENTS

4.1 Introduction..49
4.2 History of Electric Cars..51
4.3 Overview of Electric Vehicles...54
    4.3.1 Product Characteristics..54
    4.3.2 Main Reasons for the Rising Demand for Electric
        Vehicles...55
4.4 Challenges in Mass-Marketing Electric Vehicles.............59
4.5 Electric Vehicle Total Ownership Cost Advantages 62
    4.5.1 Vehicle Operating Costs..62
    4.5.2 Vehicle Maintenance Costs....................................64
4.6 Three Selected Case Studies of EVs Manufactured in Japan 65
    4.6.1 Case Study 1: Subaru Stella Plug-in....................65
    4.6.2 Case Study 2: Mitsubishi i-MiEV 67
    4.6.3 Case Study 3: Nissan Leaf....................................72
4.7 Conclusion 75

4.1 INTRODUCTION

On May 19, 2009, the *Nikkei Shimbun*, the major Japanese economic daily newspaper, dedicated three articles to the future of that country's car industry. The overall message was that the time for alternative energy vehicles had arrived. Honda had released the second-generation version of its hybrid car, the Insight, in February. Toyota had just launched the third generation of its hybrid offering, the Prius (Honda 2008; Nikkei Shimbun 2009b; Toyota 2009a). The Japanese government provided further stimulus to general consumption through the introduction of a clunkers program that resulted in skyrocketing demand for the Prius, making it the top-selling car in Japan during the ensuing 6 months (Mainichi Shimbun 2009).

This chapter examines trends in electric powertrains in the Japanese car industry in early 2010 as hybrid models gained mass market appeal, and focuses specifically on the evolution of plug-in electric cars. Careful assessment of the Japanese car market indicates that battery-powered electric cars will penetrate the world market much more quickly and deeply than generally expected.

In recent years, Japanese automotive manufacturers have established a global presence as the world's major mass producers of hybrid vehicles (HVs) essentially by successfully marketing the two models mentioned above. In March 2009, Toyota announced it had sold its 1 millionth HV worldwide (Toyota 2009b). These vehicles, demand for which had initially been restricted to environmentally conscious enthusiasts and celebrity stars, had undoubtedly reached the mainstream. The increase in crude oil prices, especially between 2006 and 2008, together with the increasingly affordable price tags of these vehicles, rendered HVs a viable alternative economic solution for the average consumer in Japan, in the United States, and elsewhere.

Both Toyota and Honda priced their new HV models at historical minimum levels. In late 2009 Japanese customers could buy a Toyota Prius for as low as ¥2,050,000 or a Honda Insight for ¥1,890,000 (respectively roughly US$21,800 and US$20,140 at the 4/30/2010 exchange rate of 93.85 Japanese yen per U.S. dollar). This immediately took the wind out of the sails of hardcore HV technology skeptics by exposing the fallacy of one of their oft-repeated arguments, namely that a necessary condition for a further boost in HV sales was another significant increase in oil prices. Even more importantly for the purposes of this chapter, as of mid-2009 the Japanese media were increasingly shifting their attention toward another eco-friendly personal transportation segment: electric cars. In that period the press was flooded with a new wave of zero-emission vehicle development stories, most of which seemed to focus on the battery-powered electric vehicles (EVs). Several different manufacturers were making headlines during this time. In June 2009, Fuji Heavy Industries (FHI) introduced a new battery-powered EV, using a mini-vehicle body assembly that was already in production, the Subaru Stella Plug-in (Japanese Road Transportation Vehicle Law defines mini-vehicle as a car within the dimensions of 1.48 m width, 3.40 m length, and 2.00 m height using an engine with displacement size under 660 cm^3). This was followed by Mitsubishi Motors Corporation (MMC), which officially announced the launch of its mass-marketed EV, i-MiEV. Last but not least, Japan's third largest automotive manufacturer, Nissan, began publicizing its intent to start production and sales of its own zero-emission EV—the Nissan Leaf—in the second half of 2010 (Nikkei Trendy Net 2009).

This chapter positions EV technology and examines the desirability of zero-emission vehicles. The criteria considered include an examination of the overall characteristics of EVs and their market readiness, including a summarized cost-benefit analysis from the point of view of the consumer. Additionally, this chapter addresses remaining challenges for full EV adoption, including necessary infrastructure improvements, vehicle sales prices, and other issues. These and other factors require further resolution if EVs are to become a lasting viable alternative in customers' minds. Finally, the chapter focuses more closely on recent EV commercial introductions by Mitsubishi and FHI (Subaru 2009), as well as the announced introduction of the Nissan Leaf by examining them as case studies.

4.2 HISTORY OF ELECTRIC CARS

A brief overview of the history of EVs provides added perspective to current events, suggesting that, contrary to popular belief, the electric car is not the product of recent technological developments. Electric vehicles first appeared more than one century ago both as a concept and as a product. The first electric cars were already operational in the latter part of the 19th century. The first actual EV, powered by a lead acid-based battery, was assembled in 1873 and sold in 1888 (Westbrook 2001, 9). Therefore, electric cars were invented, developed, and introduced to the market slightly before the gasoline–internal combustion engine-powered automobile which, according to some authors, was invented by Gottlieb Daimler in 1885. (Other sources point to Karl Benz as the inventor of the combustion engine automobile in the same year of 1885.) The combustion engine automobile was introduced commercially shortly thereafter. Electric cars were also the first type of vehicle to reach a speed of

100 km/h (62.5 mph). Their more stable performance when compared to gasoline cars secured them about 40% of the total market share of automobiles in the United States in the first years of the 20th century. Their overall superiority as a purchase option for customers in the marketplace lasted until the 1920s when they were overtaken by gasoline engine vehicles (Westbrook 2001, 19).

There were three main causes for combustion engine vehicles' emerging dominance in automobile propulsion technology and clear majority market share by 1920. First, Henry Ford developed the assembly line to manufacture his combustion engine vehicle, making mass production possible and thereby achieving economies of scale that enabled significant price reductions. The Ford Motor Company became the leading automobile manufacturer in the United States. Second, the discovery of oil reserves in Texas allowed for the necessary increase in overall domestic gasoline production and brought down the price of this commodity, which was essential for popularizing the combustion engine. Lastly, because the U.S. energy generation and supply infrastructure was still under development, gasoline was a preferred option for the not yet electrified rural areas, where it was easier to distribute gasoline for use not only in personal transportation but also for individual-scale electric power generation.

For nearly 70 years following the early 1920s, electric cars were reduced to a stopgap solution, appearing as a temporary alternative for personal transportation in times of insufficient fossil energy supply, as in World War II. Partly due to increased demand from military customers throughout most of the 20th century, automobile developers improved the reliability and durability of gasoline engine vehicles dramatically. In the post-war period, there were several attempts to develop a newer generation of EVs; however, most of them ended as mere evaluations of prototype models, with not one EV making it to the market-testing phase of trial by the general customer.

In 1996, General Motors launched the EV1, a lead acid battery-powered vehicle, the first modern-day mass-produced EV. In 1998 GM introduced an improved model with a nickel-metal hydride (NiMH) battery, anti-lock brakes, dual airbags, and additional amenities such as a stereo CD player and cruise control. It was made available to the public through a leasing agreement directly between the manufacturer and the end user. Unfortunately, perhaps due to a lack of broader institutional support, its production was stopped in April 2000 and GM terminated lease contracts without any further clarification. These circumstances were examined in a documentary ("Who Killed the Electric Car?") that raised several hypotheses, including the influence of Big Oil and its close relationship with the George W. Bush administration, which might have played a part in the demise of the promising project (Paine 2006).

More recent attempts to develop a mass-produced EV in the United States seem to have had stronger institutional (as well as consumer) support. Tesla Motors introduced a prototype of an electric luxury sports car in July 2006 and eventually began general production of its Tesla Roadster model in 2008 (Tesla 2008). This two-seater vehicle cost around $100,000 in 2009, making it an option only for the wealthiest consumers. In October 2009, Tesla announced that it would introduce a battery-powered sedan, the Tesla Model S, designed for four passengers and carrying a price tag of less than $50,000. Other publicized alternative powertrain solutions, such as the plug-in hybrid Chevrolet Volt or the EV version of BMW's e-Mini, were in the conceptual stage at this writing and represented promise of stiff competition to Japanese EV producers in years ahead as EV technology developed further.

EV powertrain technology was dominated until recently by American automotive manufacturers. However, at the end of 2009 there was no doubt that the most advanced mass-marketable solutions for personal transportation using battery-powered EVs were to be found in Japan, mostly due to the combination of manufacturing and technological sophistication, institutional support, and consumer readiness. During the summer of 2009, both MMC and Subaru (FHI) introduced their newest generation of EVs to the Japanese market; both models were based on pre-existing mini-vehicles. These EVs were expected to be sold initially to corporate clients and local government authorities. By the end of 2009, both companies were planning to launch sales to individual customers in 2010. In the summer of 2009, Nissan announced that it would start to produce and sell a brand-new model of electric car, named Leaf, in the second half of 2010 in Japan, Europe, and the United States. Toyota also announced the future launch of EVs under development in 2012. On May 20, 2010, Toyota Motor Corporation (Tesla 2010) announced its intention to work jointly with Tesla Motors on EV development. At the same time, Toyota purchased a US$50 million stake of Tesla common stock.

4.3 OVERVIEW OF ELECTRIC VEHICLES

4.3.1 Product Characteristics

Generally speaking, the concept of EVs is very straightforward. The EV powertrain has three main components: an electric motor, an electric battery for energy storage, and a power management unit.

The vehicle is solely propelled by an electric motor with either direct current (DC) or alternating current (AC). DC-type motors were used in most vehicles built until the 1990s. Ease of control and a generally high level of development made them the most common choice. In the 1990s, the AC motor, which has smaller dimensions and requires less maintenance, evolved through focused R&D efforts and started to replace the DC motor. The more recent EVs and HVs employ AC and brushless motors, which can be induction motors, permanent magnet motors, or switched reluctance motors (Husain 2003, 95).

The battery is charged by a source of electricity, either in the form of a regular household socket or of a fast-charging station. Battery-related specific performance and behavior depend on the battery's chemical composition. The materials used in most batteries are NiMH and lithium-ion. The NiMH battery, which is also used in the Toyota Prius, has similar characteristics to that of the previously widespread nickel-cadmium type. The major difference is that instead of cadmium, which is considered environmentally hazardous, it uses a hydrogen-absorbing metal alloy as a negative electrode, with nickel oxyhydroxide functioning as the positive electrode (Larminie and Lowry 2003, 38). The lithium-ion battery can achieve higher power and energy levels than an NiMH battery pack, but it has higher production costs. It has a weight advantage over other battery technology types, which makes it a great candidate for EVs, because it can offer up to three times the specific energy level of lead acid batteries. The lithium-ion battery uses a lithiated transition metal intercalation oxide for anode and lithiated carbon for cathode and requires a liquid solution or a solid polymer as an electrolyte (Larminie and Lowry 2003, 45). The planned introduction of EVs to the marketplace is yet another application of the lithium-ion battery technology, which has been accelerated by the recent demand for notebook computers, portable music players, and mobile phones.

The power management unit is the electronic brain of an EV. It not only ensures that optimal power is supplied to the motor while controlling the overall vehicle performance (serving as a standard vehicle electronic control unit) but also oversees the recharging of the battery by the generator through recouping kinetic energy when the vehicle's brakes are applied and converting this kinetic energy to electricity.

4.3.2 Main Reasons for the Rising Demand for Electric Vehicles

EVs can be classified as zero-emission vehicles because they do not produce any exhaust gases or air pollutants (CARB 2009). There is no combustion involved in vehicle propulsion, so EVs have virtually no road noise except for tire roll sound. As such, EVs are a proven "strategy to de-carbonize transport and improve air quality in highly developed urban areas" (Mayor of London 2009).

The recent increase in public interest in EVs as a green transportation alternative to mitigate global warming has been found to be linked to five separate focal areas of consumer perception: energy sources and supply reliability, greenhouse gas emission limits and environmental regulation, cost reductions for the consumer over the EV product life, improved vehicle technology, and adoption of a more sustainable lifestyle (Mochimaru and Lee 2009, 9; Electrification Coalition 2009, 10). Since the beginning of the Obama administration in January 2009, the United States (the world's biggest consumer) has begun implementing initiatives to break away from domestic dependence on fossil fuels and steering the nation to contribute to the international effort to resist global warming.

As depicted in Figures 4.1, 4.2, and 4.3—which compare energy consumption, CO_2 emissions, and expected powertrain technology dissemination in Japan with other regions of the world—transportation represents about 23% of all annual worldwide CO_2 emissions and 19% of all emissions in Japan, where about 48% of the transportation portion is composed of emissions produced by personal vehicles (IEA 2009; MEJ 2008). This places EVs high on the priority list in the Japanese national effort to reduce overall emissions by a targeted 25%. The following are expected benefits to be derived from widespread EV use:

- Reduced dependence on fluctuating fossil energy prices
- Reduced dependence on imported oil (annual capital outflows, national security)

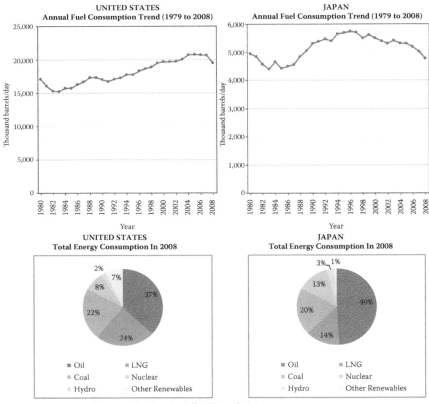

Source: Energy Information Administration. Official Energy Statistics.

FIGURE 4.1

Oil and energy consumption in the United States and Japan. (International Energy Agency, *World Energy Outlook 2009,* OECD/IEA, 2009.)

- Reduced greenhouse gas emissions, seeking to meet national targets
- Stronger environmental regulations, especially in the European market
- Reduced noise pollution in heavily populated urban areas (EVs have virtually no running noise, only tire roll noise)
- Revised policies by central and local governments (clean-tech regulatory framework)
- Reduced total cost of ownership (purchase, operation, and maintenance of EV)
- Increased public awareness and involvement in ecological impact of transportation
- Major advances in battery technology (possible cost reduction)

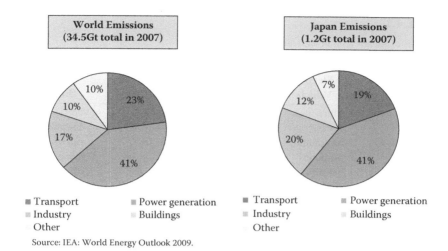

FIGURE 4.2
CO_2 emissions in Japan and the rest of the world. (International Energy Agency, *World Energy Outlook 2009*, OECD/IEA, 2009.)

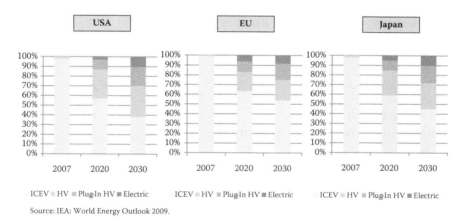

FIGURE 4.3
Expected powertrain technology diffusion in the United States, European Union, and Japan. (International Energy Agency, *World Energy Outlook 2009*, OECD/IEA, 2009.)

Current HVs offer the possibility of driving only very short distances on purely electric power. Therefore, in order to make a rational comparison among the alternative power and propulsion options for personal vehicles, the list in Table 4.1 is restricted to plug-in hybrid, flex-fuel, and fuel-cell technologies. Despite their excellent ecological performance, EVs have three main negatives that make them less

TABLE 4.1

Analysis of Clean-Tech Powertrain Technologies

	Electric Vehicle	Plug-in Hybrid	Fuel-Cell Vehicle	Flex-Fuel Vehicle (Bio-Fuel)
	Vehicle uses electric motor and on-board battery charged from external source	Vehicle can run on electric motor drive or IE CE together The battery can be charged externally	Vehicle propelled by electric motor Fuel cells inside the vehicle create electricity using hydrogen and oxygen	Vehicles designed to run on ICE but using a blend composed of gasoline and up to 85% of ethanol
Energy Storage	Battery	Gasoline tank battery	Hydrogen tank battery	Bio-fuel tank
Vehicle Propulsion	Motor	Engine motor	Fuel cell motor	Engine
Air Pollutants Reduction	A	B to A	A	B
CO_2 Emission Reduction	A	B to A	A	B to A
Oil Dependency Reduction	A	B to A	B to A	A
Fuel Availability	B	B	C	C
Tanking/Charge Time	D	Battery = C Tank = B	B	B
Operating Distance	C to D	A	C to B	C to B
Vehicle Affordability	C	B	D	B

Legend: A = Excellent; B = Good; C = Average; D = Poor
Evaluation Standard: Gasoline vehicle = B
Source: Kawai and Niikuri (2009).

attractive when compared to the other three competing green powertrain technologies. EV technology lags behind (1) in the maximum distance the vehicle can travel between refueling stops, (2) in total time needed to charge/refuel the vehicle, and (3) in the price burden at the time of purchase.

4.4 CHALLENGES IN MASS-MARKETING ELECTRIC VEHICLES

It follows that the three key issues that need to be resolved for EVs to become a viable purchase option for consumers are travel distance between charges, infrastructure of existing charging stations, and vehicle price burden to the customer.

First, per-charging distances need to improve so consumers can travel up to medium ranges without the risk of running out of power. Increased distance per charging can be achieved either through the development of new materials and structures to allow for electricity storage in cells, or through the optimization of the power management system that allocates the consumption of electricity stored onboard. On average, EVs have 10,000 assembly components, roughly a third of internal combustion engine vehicles. Furthermore, they have only one moving part in the motor, no engine cooling components, no gearbox, and no multi-speed transmission. As a result, the electric motor itself is very standardized and available off the shelf. The most sophisticated developmental concern is the achievement of a balance between energy storage needs and distance traveled requirements; that is, the tradeoff is between the weight of energy-storing devices (batteries) and the autonomy that is sought.

Second, the current lack of battery-charging stations presents an infrastructure constraint that needs to be overcome. This could be achieved through coalitions and/or cooperative initiatives with governments and/or other industries. In comparison to standard internal combustion engine vehicles, which operate on average for 500 to 600 kilometers (km) with a full tank, EVs can be driven up to roughly 200 to 300 km per full charge. Under these circumstances, customers will find themselves in need of recharging twice as frequently as they now need to refuel, which is not unreasonable since, for example, surveys in Japan also show that 93% of individual mini-vehicle drivers and 88% of corporate mini-vehicle drivers drive an average of 40 km or less on a daily basis (Barclays 2009, 9). There also is the issue of necessary time to recharge. Consumers will be able to use a standard household electricity plug to recharge their vehicles. However, using a 100V power outlet, it can take anywhere from 8 to 14 hours to fully recharge (standard household power outlets in Japan use voltage of 100V, unlike the 110V that is standard in the United States and Canada). If the household has or is able to obtain access to a 200V power source, this time can be reduced by half. In Japan this is a rather rare alternative since it would entail access to an industrial electricity installation.

Consequently recharging the battery would not seem to be a problem for a user who makes routine trips in the same area or uses the vehicle to commute to and from work, but it would become undesirable for uneven patterns of vehicle use, because of the need to recharge the EV while away from home. Therefore publicly available fast-charging stations are a necessity. These fast-charging facilities would operate at a higher voltage level, usually through a three-phase circuit and would use a vehicle-equipped inverter that would allow for battery charging at high voltage levels. As these charging docking stations represent a significant investment, they should be located in areas of high expected demand, such as commercial centers or shopping malls. Additionally, since a three-phase circuit socket is used (a nationwide standard), certain cost synergies can be realized with plug-in hybrid facilities.

Third, vehicle prices could be high for the average consumer. Original equipment manufacturers (OEMs) need to be given governmental support, for example through tax incentives or through other forms of subsidy, in order to reduce manufacturing costs and consequently final customer prices. Both the Subaru and Mitsubishi EV models cost approximately ¥4,500,000, which is equivalent to a higher-end medium-sized gasoline-powered vehicle. The Japanese government offers some subsidies for the purchasers of these vehicles. By sending an application to the Center for the Next Generation Vehicle Promotion, individual buyers can be eligible for a subsidy of up to ¥1,380,000. Also, the Japanese Ministry of Land and Transportation offers additional tax rebates on the car weight tax (¥13,200) and on the car acquisition tax (¥121,500). Additionally, there are local government subsidies and incentives for small businesses in Japan. For example, the Tokyo metropolitan government provides a ¥770,000 subsidy to small business owners purchasing an EV. Taken in aggregate these subsidies reduce the difference in price between a conventional internal combustion engine vehicle (ICEV) and an EV by 50%.

Economies of scale due to increased production volume of mass market EVs are expected to help make further price reductions possible. In late 2009 the lithium-ion battery pack represented roughly 50% of the total vehicle purchase price. The industry-wide average price per battery output was $600 per kWh, regardless of the actual battery production volume, its chemistry composition, the vehicle type, or the battery pack size (Electrification Coalition 2009, 74). Applying this average price to the two cases of EVs for which detailed information is available, an estimated price of $9,600 and $5,400 respectively for the Mitsubishi and the Subaru vehicle battery packs is obtained, based on the vehicles' nominal electric power, although the actual prices of the battery packs were expected to be higher. In the case of the i-MiEV, the price of a lithium-ion battery pack could be assumed to reach anywhere from 30% to 50% of the total price of the EV, leading to an overall price range from $14,000 to $23,000.

The spring 2010 Nissan pricing announcement for the United States is also very competitive, setting the manufacturer's suggested retail price (MSRP) at $32,720. If the cash-for-clunkers U.S. government-sponsored incentive of $7,500 were to be applied, the final price would hypothetically drop to $25,280. Nissan also publicly stated that it would lease the Leaf for $349 a month for 36 months after a $1,999 down payment is made (Nissan 2010a). This brought the premium paid for a zero-emission vehicle down even further, much closer to the price tags of hybrid powered vehicles mentioned in the introduction, potentially increasing the pool of buyers who could afford an EV.

Better Place, a California-based company started in 2007 by Shai Agassi, an Israeli entrepreneur, has aimed to provide a solution to help overcome fossil fuel dependency in the form of charging stations for EVs. The company's corporate mission is to tackle the second and the third constraints mentioned above. In order to facilitate the adoption of EVs on a global scale, this company specializes in providing electric charging infrastructure, which includes regular charging and battery-swapping stations that can be used if a driver needs to travel more than the given maximum driving radius. The business concept is innovative, positioning Better Place as the owner of the batteries, which can be exchanged easily due to their modular design implemented through cooperative development with automakers. The EV user would pay a monthly usage fee for consumed electricity and access to this chain of battery-charging stations. At this writing Renault has already announced its E.Z. Concept, which it intends to make compatible with Better Place's charge system (Renault 2010). Furthermore, at the time of writing, Better Place had begun building stations to test its concept in Israel and had additional systemic tests planned for Denmark and Hawaii.

The development of EVs requires the combined expertise and cooperation of automotive and electronics manufacturers. A Boston Consulting Group (BCG) report on the comeback of EVs mentions an additional important factor for this development. OEMs should strategically build new supply chains and "be pursuing partnerships with battery suppliers, including, for example, those between Toyota and Panasonic, Volkswagen and Sanyo, Bosch and Samsung, and Renault-Nissan and NEC" (Book et al. 2009, 8). Until recently the vehicle powertrain development was mostly in the hands of research and development arms of OEMs. This was true despite the fact that collaboration between OEMs and electronics parts suppliers was increasing as a result of recent growing usage of electronic control components in vehicles. The EV powertrain is composed of much less mechanical hardware, and electronics play a major role in determining the actual vehicle performance and reliability.

4.5 ELECTRIC VEHICLE TOTAL OWNERSHIP COST ADVANTAGES

Despite these challenges there is a strong mitigating factor when consumers consider the EV option: the expectation that the total cost of ownership (TCO) is competitive with that of conventional cars. In order to enlighten the general public on the economic benefits of operating EVs, OEMs need to develop more focused informational and promotional initiatives to make sure these benefits are fully understood. The estimate of total cost of ownership informs how much a vehicle would cost a customer throughout its product life. Most of the EVs available in the marketplace have a high purchase price. This represents a large upfront investment for an individual and represents a difficult hurdle for customers to overcome in justifying their purchase. When lifetime fuel consumption and vehicle maintenance costs are included in the analysis, however, EVs are found to be competitive.

4.5.1 Vehicle Operating Costs

Information provided on the governmental Internet portal of the Energy Information Agency can be used to calculate vehicle operating costs. The site offers information regarding the average price of a gallon of gasoline and 1kWh of electricity in the United States (specifically in the state of Pennsylvania). In order to understand how prices are different in Japan, websites of major utility companies were consulted. Prices cited are for electricity use at the higher tariff charged during the day and neglect discounts available to households either for using cheap night current or for consuming larger monthly amounts of electricity. Table 4.2 shows results of a simulation of operating costs in U.S. dollars per 100 miles.

TABLE 4.2

Automobile Operation Costs

Operation Costs in USD per 100 Miles: Energy Cost Table

Power	Price (Local)	Price (USD)	Per Unit	Explanation
Gasoline (PA)	$ 2.9626	$2.9626	/Gallon	Average price in State of Pennsylvania (Note 1)
Gasoline (US)	$ 2.6390	$2.6390	/Gallon	Average retail price in United States (Note 1)
Gasoline (JPN)	¥ 423.97	$3.9938	/Gallon	Average retail price in Tokyo (Note 2)
Electricity (PA)	$ 0.1244	$0.1244	/KWh	Residential price in State of Pennsylvania (Note 3)
Electricity (US)	$ 0.1205	$0.1205	/KWh	Average residential price in USA (Note 3)
Electricity (JPN)	$ 22.86	$0.2153	/KWh	Average residential price in Japan (Note 4)
JPY100/USD	$ 0.942	As of April 06, 2010		

Simulation Based on United States and Japan Average Prices of Electricity and Gasoline

Power	Consumption	USA	Japan	Remarks
Gasoline	25 MPG	$10.56	$15.98	
	35.5 MPG	$ 7.43	$11.25	Based on Obama administration CAFÉ 2016 standard
Electricity	125 Wh/Km	$ 2.43	$ 4.33	Mitsubishi i-Miev: 20.125K Wh/100 Miles
	150 Wh/Km	$ 2.89	$ 5.17	Nissan Leaf Simulation: 24KWh/100 Miles
Plug-In Hybrid	40 Miles on EV	$ 0.97	$ 1.73	Used Nissan Leaf electricity consumption
	60 Miles on Gas	$ 3.17	$ 4.79	Used Toyota Prius 50MPG catalog fuel consumption
	Total (EV + Gas)	$ 4.14	$ 6.53	

Sources:

Note 1: Energy Information Administration. Retail Gasoline Prices By Region (http://www.eia.doe. gov/oil_gas/petroleum/data_publications/wrgp/padd_1b_mini_report.html)

Note 2: Current Gasoline Retail Prices List in Japan (http://gogo.gs/rank/13.html)

Note 3: Energy Information Administration. Average Retail Price of Electricity (http://www.eia.doe. gov/cneaf/electricity/epm/table5_6_a.html)

Note 4: Tokyo Power Company: Charge for consumption monthly over 120K Wh (http://www.tepco. co.jp/e-rates/individual/fuel/adjust/index-j.html)

Results show that EVs are 1.43 times cheaper to operate than plug-in HVs (according to expected fuel economy specifications announced by automakers) in the United States and 1.26 times cheaper in Japan. If one were to compare this to the new CAFÉ 2016 Standard, in which the Obama administration sets the average fuel consumption to 35.5mpg, one would find that EVs are 2.57 times more economical than ICEVs in the United States and 2.18 times more economical in Japan as a solution for personal transport. Only very few gasoline vehicles available now are actually capable of providing such fuel efficiency.

Some studies state that EVs can run for as little as $0.01 per kilometer, making this option even more attractive. For example, the Nissan Leaf preorder sales launch press release from March 30, 2010, included Nissan's cost of ownership calculation for the EV of ¥86,000 for a period of six years in comparison to the cost of a regular ICEV of ¥670,000, when used to travel 1,000 km each month and recharged at the Tokyo nighttime electricity tariff (Nissan 2010b).

4.5.2 Vehicle Maintenance Costs

EVs have fewer components than their IECV counterparts (a large reduction in assembly parts is possible because of the inexistence of a combustion engine, of engine-related cooling parts, of transmission, and of a gearbox). HVs need even more parts to complete the vehicle assembly, because besides the standard combustion engine, the vehicle also carries an electric motor, a battery, an inverter, and additional electronic parts.

The biggest reduction in number of parts is apparent in the powertrain technology. A standard gasoline-powered car engine has a few dozen moving parts. In contrast, the electric motor has only one. Moving parts are subjected to the most friction and wear-and-tear resulting in higher maintenance needs. These components require regular replacement to secure smooth vehicle functioning. EVs also do not need liquid oil and filter or transmission fluid changes. This fact is extremely important to consider, as a potential buyer can expect higher cash outflows related to the maintenance of ICEVs or even more sophisticated HVs. Maintenance costs for either of these options usually total about $500 per year for the first five years, more than double the costs for EVs.

4.6 THREE SELECTED CASE STUDIES OF EVS MANUFACTURED IN JAPAN

What follows is a discussion of three EVs developed in Japan which were closest to market in early 2010. The information contained herein was obtained from the sources cited as well as on-the-ground primary research in Japan.

4.6.1 Case Study 1: Subaru Stella Plug-in

The Subaru Stella Plug-in is based on a mass-produced 660 ccm gasoline mini-vehicle. The Stella Plug-in is designed to be a city commuter vehicle requiring only a short amount of time to recharge before operation. It can operate up to approximately 90 km on full charge. When charged by a fast-charger for 15 minutes, it can travel up to 70 km. The basic idea behind this concept was to create a well-balanced EV, which could be conveniently charged while also providing a decent operating range to support a customer's daily commute. If recharged at home, the vehicle needs 5 hours using 200V, or eight hours using 100V. These specifications make the plug-in Stella a great choice for customers who are running their errands within an urban or suburban area or require a vehicle to commute to work daily. Table 4.3 presents the main specifications of the Subaru Stella Plug-in.

The installed electric motor can produce 170 Nm of torque despite its output of 47 kW, which is the maximum power output for 660 ccm gasoline mini-vehicles. The motor is located in the front of the vehicle, the lithium-ion battery pack is hosted below the floor, and the vehicle can accommodate up to four passengers. Figures 4.4 through 4.8 illustrate the main characteristics of this automobile. According to Subaru's press release, the plug-in Stella was made available to general users in June 2009, and 170 vehicles had been delivered by end of March 2010. Subaru planned to produce and sell a total of 200 vehicles in the fiscal year beginning April 2010. FHI's press release mentioned that the manufacturer pursued "the perfect integration of pleasant and reliable driving with environmental considerations in its vehicle development philosophy. In its efforts to constantly enhance its current power unit lineup, FHI positions its EVs as one of the viable solutions and key technologies for environmental preservation, and it will further research and develop its EVs" (Subaru 2009).

TABLE 4.3

Subaru Stella Plug-In Vehicle Information

Specifications	
Length × Width × Height	3,395 mm × 1,475 mm × 1,660 mm
Vehicle weight	1,010 kg
Number of passengers	4
Maximum speed	100 km/h
10•15 Consumption mode (AC)	105 Wh/km
10•15 Consumption mode maximum travel distance	90 km
Maximum power	47 kW
Maximum torque	170 N•m
Drivetrain	Front wheel drive
Vehicle price including 5% consumption tax (JPY)	¥4,725,000
Speed charging to 80% (200V-50 kW)	15 minutes
Household socket 200V (15A) to 100%	5 hours
Household socket 100V (15A) to 100%	8 hours
Sales launched	June 2009

Source: http://www.subaru.jp.

FIGURE 4.4

Subaru Stella Plug-in: vehicle appearance. (Fuji Heavy Industries (Subaru) website. With permission.)

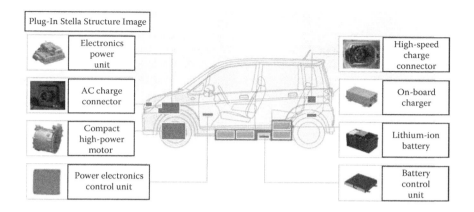

Plug-In Stella Structure Image

Electronics power unit

AC charge connector

Compact high-power motor

Power electronics control unit

High-speed charge connector

On-board charger

Lithium-ion battery

Battery control unit

FIGURE 4.5
Subaru Stella Plug-in: powertrain layout. (Fuji Heavy Industries (Subaru) website. With permission.)

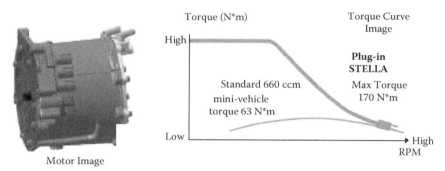

Torque (N*m)

Torque Curve Image

High

Plug-in STELLA

Standard 660 ccm mini-vehicle torque 63 N*m

Max Torque 170 N*m

Low

High RPM

Motor Image

Source: Fuji Heavy Industries (Subaru) website.

FIGURE 4.6
Subaru Stella Plug-in: permanent magnet synchronous motor with performance chart. Oil and energy consumption in the United States and Japan. (Fuji Heavy Industries (Subaru) website. With permission.)

4.6.2 Case Study 2: Mitsubishi i-MiEV

The Mitsubishi i-MiEV has a 2,550 mm long wheelbase and hosts 230 kg of high-capacity lithium-ion battery packs below its floor. The vehicle is built to accommodate four passengers. The i-MiEV has an electric motor of 47 kW (64 BHP) output, which is equivalent to the gasoline turbocharged original mini-vehicle and is comparable to the aforementioned Subaru model. The motor's torque is almost twice as large and can reach about 180 Nm. The vehicle can be recharged in two ways. If an infrastructure

On-board battery is composed of 16 modules below floor. As shown on the diagram, 8 modules are located in the front and 8 in the rear of the vehicle.

Source: Fuji Heavy Industries (Subaru) website.

FIGURE 4.7
Subaru Stella Plug-in: on-board high-capacity lithium-ion battery. (Fuji Heavy Industries (Subaru) website. With permission.)

FIGURE 4.8
Subaru Stella Plug-in: vehicle charging station. (Fuji Heavy Industries (Subaru) website. With permission.)

fast-charger is used, the vehicle's batteries can be recharged to 80% of their capacity in 30 minutes. In the case of a regular household socket, the vehicle can be fully recharged in either 7 hours using 200V or 14 hours using 100V. Mitsubishi reported that the initial batch of fleet orders was placed by the Tokyo Electric Company and by Japan Post Holdings. In order to supply the vehicles with lithium-ion batteries, Mitsubishi (8.3%) formed

TABLE 4.4

Mitsubishi i-MiEV Vehicle Information

Specifications	
Length × Width × Height	3,395 mm × 1,475 mm × 1,610 mm
Vehicle weight	1,100 kg
Number of passengers	4
Maximum speed	130 km/h
10•15 Consumption mode (AC)	125 Wh/km
10•15 Consumption mode maximum travel distance	160 km
Maximum power	47 kW
Maximum torque	180 N•m
Drivetrain	Rear wheel drive
Vehicle price including 5% consumption tax (JPY)	¥3,980,000
Speed charging to 80% (200V-50 kW)	30 minutes
Household socket 200V (15A) to 100%	7 hours
Household socket 100V (15A) to 100%	14 hours
Sales launched	July 2009

Source: http://www.mitsubishi-motors.co.jp. With permission.

a joint venture, named Lithium Energy Japan (LEJ), with GS Yuasa (51%) and the Mitsubishi Corporation (40.7%). Mitsubishi will be able to begin mass production of enough lithium-ion batteries for 50,000 EVs per year from 2012 (Mitsubishi 2010). LEJ plans to expand this capacity in the future to supply batteries for up to 100,000 EVs. Table 4.4 describes the main specifications of the Mitsubishi i-MiEV and Figures 4.9 through 4.13 illustrate the main characteristics of this automobile.

In June 2009, Mitsubishi started delivering a first batch of 1,400 vehicles to corporate customers with an initial price of ¥4,599,000. Mitsubishi Motors CEO Osamu Masuko, in an interview with *Nikkan Kogyo Shimbun*, the Japanese daily industrial newspaper, on July 28, 2009, stated the following:

[T]he vehicle production is constrained by battery pack production capacity, which will limit the volume to 6,000 vehicles in 2010. In 2011 Mitsubishi expects to manufacture 15,000 units, eventually ramping up its output to 30,000 vehicles in 2013. From the second half of 2010, a left-hand drive option will be introduced, as the plan is to export 1,000 vehicles during that year. We already have requests from various countries and local governments to receive these vehicles. There are many purchase orders and we

FIGURE 4.9
Mitsubishi i-MiEV: vehicle appearance. (Mitsubishi Motors Corporation website. With permission.)

> need to catch up with them... This project is part of our corporate social responsibility and our contribution to the effort to create a low carbon emissions society. (Nikkan Kogyo Shimbun 2009)

In March 2009, Mitsubishi announced that it agreed to design a version of the i-MiEV for the European market in partnership with the PSA Group. The two car manufacturers planned to launch these units under the Mitsubishi (Mitsubishi 2009), the Citroen model C-Zero, and the Peugeot model Ion (PSA 2009) brands by the end of 2010. On December 2, 2009, the PSA Group announced its intent to become the largest shareholder of

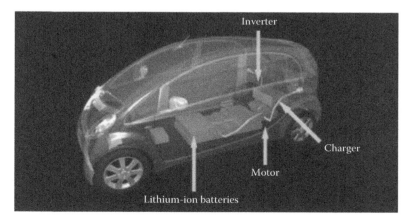

FIGURE 4.10
Mitsubishi i-MiEV: powertrain layout. (Mitsubishi Motors Corporation website. With permission.)

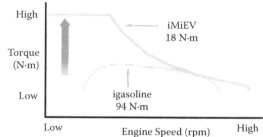

Permanent magnet synchronous motor

High

Torque (N·m)

Low

Low

iMiEV 18 N·m

igasoline 94 N·m

Engine Speed (rpm)   High

FIGURE 4.11
Mitsubishi i-MiEV: permanent magnet synchronous motor and performance chart. The above electric motor, which is assembled into i-MiEV, has only one moving part. The graph on the right shows the performance of the motor, when compared to standard 660ccm gasoline combustion engine with turbocharger Mitsubishi "i" mini-vehicle. (Mitsubishi Motors Corporation website. With permission.)

MMC through the purchase of a ¥200 to ¥300 billion stake in the company and was finalizing negotiations to acquire 30% to 50% of voting shares (Nikkei Shimbun 2009a). However, on January 25, 2010, the intent to purchase these Mitsubishi shares was rejected by the Peugeot family, PSA's key shareholder. That decision came on the heels of the global economic crisis, which unfavorably impacted PSA's market valuation and led PSA to choose to pursue a strategic partnership with joint projects instead, including the electric car (Reuters 2010). This looser alliance could reduce knowledge transfer between the two companies but was not expected to affect the joint effort to globally introduce the i-MiEV.

On March 30, 2010, Mitsubishi (Mitsubishi 2010) announced that it had lowered its manufacturer's suggested retail price further by ¥619,000 to ¥3,980,000. This meant that if the Japanese government continued to

Battery pack

Module

FIGURE 4.12
Mitsubishi i-MiEV: on-board high-capacity lithium-ion battery. Lithium-ion battery, which is located below the floor of the vehicle, contains high-capacity 88 50Ah cells connected in a series. Battery module consists of 4 cells, and therefore on-board battery contains 22 modules. (Mitsubishi Motors Corporation website. With permission.)

FIGURE 4.13
Mitsubishi i-MiEV: vehicle at charging station. (Mitsubishi Motors Corporation website. With permission.)

extend the same incentives as they did in 2009, customers would be eligible for the same ¥1,140,000 subsidy, lowering the final price for customers to ¥2,840,000.

4.6.3 Case Study 3: Nissan Leaf

There are two main differences distinguishing the Nissan Leaf from the two automobiles profiled above. First, Nissan designed its EV as a major change to a current model (Tiida) instead of choosing to modify only the powertrain of an existing model. Second, the vehicle has the dimensions of a regular passenger car, and as such it is much larger in size than the two previously discussed vehicles.

Nissan unveiled the Leaf in early August 2009. Nissan's official stance regarding eco-friendly vehicles has been one of skepticism toward the mass production of HVs. The company has been generally lukewarm about introducing hybrid technology and plans to use it only in a few high-end models, mainly the Infiniti brand. In contrast to Toyota's and Honda's promotion of HVs, Carlos Ghosn, the CEO of Renault and Nissan, strongly argued for the need for zero-emission vehicles, especially battery-powered EVs. Nissan's main argument was that a majority of users travel a maximum distance of up to 100 km daily, and therefore EVs could provide the best solution to these customers' needs, regardless of their per-battery

TABLE 4.5

Nissan Leaf Vehicle Information

Specifications	
Length × Width × Height	4,445 mm × 1,770 mm × 1,550 mm
Vehicle weight	N/A
Number of passengers	5
Maximum speed	140 km/h
10•15 Consumption mode (AC)	N/A
10•15 Consumption mode maximum travel distance	160 km
Maximum power	80 kW
Maximum torque	280 N•m
Drivetrain	Front wheel drive
Vehicle price including 5% consumption tax (JPY)	¥3,760,000
Speed charging to 80% (200V-50 kW)	30 minutes
Household socket 200V (15A) to 100%	8 hours
Household socket 100V (15A) to 100%	16 hours
Sales launched	Preorder from April 2010
	Sales from December 2010

Source: http://www.nissan.co.jp.

charge distance limitation. Table 4.5 describes the main specifications of the Nissan Leaf and Figures 4.14 through 4.17 show the actual vehicle and charging station.

The Leaf can comfortably seat up to five passengers. The electric motor can provide the vehicle with maximum power of 80 kW (approximately equivalent to a 1.5-liter gasoline engine) and supply a maximum torque of 280 Nm. The on-board lithium-ion battery pack, when fully charged, can support a maximum travel distance of 160 km. It needs the exact same time as the i-MiEV (i.e., 30 minutes) for a battery fast-charge to 80%. The car's regular household electric charging time will be up to 8 hours at 200V and up to 16 hours at 100V to fully charge. At the time of this writing Nissan had not provided any further details, but it is safe to assume that the Leaf's larger size and weight will demand a larger battery pack, which leads to the longer required charge time.

The vehicle sales launch was scheduled for December 2010. As of the beginning of April 2010, Nissan had started taking preorders for production from Japanese and U.S. customers. Nissan is in cooperation with its French ally Renault SA and aspires to be the first car

FIGURE 4.14
Nissan Leaf on display at Nissan Gallery in Ginza, Tokyo. (Photo Jan Zelezny, Nissan Gallery, Tokyo, Japan. With permission.)

manufacturer to mass-market an EV by selling to individual customers in Japan, in Europe, and in the United States. As of March 2010, Nissan had publicly announced the actual vehicle price (¥3,760,000) for which it planned to offer this vehicle (Nissan 2010b). As the same corporate press release mentioned, the Japanese government offered an incentive program which helped bring the price to Japanese residents down by ¥770,000 to ¥2,990,000. Because the Nissan Leaf is a zero-emission class vehicle, in Japan it is also exempt from car acquisition taxes and car weight taxes.

Nissan Auto Loan, the financial arm of Nissan in Japan, created an alternative purchase plan, in which a customer could make a ¥2,400,000 down payment and then would be charged ¥10,000 (including the cost of electricity) on a monthly basis to make the car a more attractive choice for potential buyers. Nissan planned to sell 6,000 Leaf units in the 2010 fiscal year.

FIGURE 4.15
Nissan Leaf front view. (Photo Jan Zelezny, Nissan Gallery, Tokyo, Japan. With permission.)

4.7 CONCLUSION

EVs are expected to be an essential part of the solution to the problem of sustainable personal transportation. As the cases described in the previous section and summarized in Table 4.6 show, automotive manufacturers such as Subaru, Mitsubishi, and Nissan are committed to contributing to this solution. As of April 2010, these automotive manufacturers already had complete products that could be ordered and delivered to their customers in only a few months' time. Electric car production and use were no longer simply figments of the imagination, but a reality that was expected to be present on our roads in the not-too-distant future. These vehicles could fulfill the needs of consumers owning multiple vehicles or who are routine commuters within a particular area seeking to reduce costs. With a travel distance of up to 160 km on full charge, these EVs would provide

FIGURE 4.16
Nissan Leaf interior. (Photo Jan Zelezny, Nissan Gallery, Tokyo, Japan. With permission.)

enough mobility for the user to handle everyday necessities. They represent a solution for a completely different segment of customers than the long-distance travelers served by hybrid vehicle and electric vehicle (HEV) or fuel-cell propulsion.

There still were a few broad issues awaiting resolution, but effective communication with the consumer was expected to mitigate any negative effects, especially as it had become more and more apparent that the zero-emissions powertrain technology could offer an effective alternative to the very entrenched system of fossil fuel–based transportation. For example, long-distance traveling still remained an issue to be resolved, and, as a result, the mass-marketing of EVs would require additional investments in infrastructure, such as building fast-charging stations. It is important to note that the essential backbone of this infrastructure was already in place with the common availability of household electricity and ready access to the power grid. Also, as we have seen in the case of Japan, to promote and stimulate the presence of EV technology in the marketplace, the institutional support of governments and local authorities was expected to be

FIGURE 4.17
Charging station on display at Nissan Gallery. (Photo Jan Zelezny, Nissan Gallery, Tokyo, Japan. With permission.)

essential. With adequate government financing, such as purchasing incentives, consumers would be more attracted to the positive externalities and overall benefits of the technology and might be more inclined to make the purchase over a standard ICEV or HV. This institutional support would also represent an additional cost which was expected to be outweighed by the positives this technology offers in de-carbonizing individual transportation on a large-scale basis.

Electric vehicles do not produce carbon dioxide emissions and they do not disseminate air pollutants, although the generation of the electricity might. Even under conservative assumptions, EVs remain superior to standard ICEVs, HVs, and the next generation of plug-in hybrids. Furthermore, electricity prices have been less volatile than crude oil

TABLE 4.6

Specifications Comparison Table

	EV Case 1	EV Case 2	EV Case 3
Detailed Vehicle Information	**Subaru**	**Mitsubishi Motors**	**Nissan**
Specification	**Subaru Stella Plug-in**	**Mitsubishi i-MiEV**	**Nissan Leaf**
Length × Width × Height	3,395 mm × 1,475 mm × 1,660 mm	3,395 mm × 1,475 mm × 1,610 mm	4,445 mm × 1,770 mm × 1,550 mm
Vehicle weight	1,010 kg	1,100 kg	N/A
Number of passengers	4	4	5
Maximum speed	100 km/h	130 km/h	140 km/h
10•15 Consumption 10•15 Consumption	105 Wh/km	125 Wh/km	N/A
Full charge travel distance	90 km	160 km	160 km
Motor type	Permanent magnetic synchronous	Permanent magnetic synchronous	Permanent magnetic synchronous
Maximum power	47 kW	47 kW	80 kW
Maximum torque	170 N•m	180 N•m	280 N•m
Drivetrain	Front wheel drive (2 WD)	Rear wheel drive (2 WD)	Front wheel drive (2 WD)
Battery type	Lithium-ion type	Lithium-ion type	Lithium-ion type
Overall voltage	346 V	330 V	N/A
Overall electric power	9k Wh	16 kWh	N/A
Vehicle price incl. 5% consumption tax	JPY 4,725,000	JPY 4,599,000	JPY 3,760,000
Speed charging to 80% (200V-50 kW)	15 minutes	30 minutes	30 minutes
Household socket 200V (15A) to 100%	5 hours	7 hours	8 hours
Household socket 100V (15A) to 100%	8 hours	14 hours	16 hours
Sales launched	Debut in June 2009	Debut in July 2009	Preorder from April 10 Debut in December 2010

Source: Individual manufacturers' websites.

prices. From 1990 to 2007, the period for which the Energy Information Administration provides data, the average price of 1 kWh of electricity in the United States has shown a fluctuation of only about $0.10 (EIA 2007). Clearly this is a much less volatile commodity than crude oil.

This chapter has mainly focused on the products of three Japanese manufacturers. With recent announcements by Toyota, Volkswagen, and other automobile manufacturers, it is becoming clear that in the years ahead, EV usage will become more widespread and therefore is expected to eventually move into the mainstream of public acceptance and use. Japanese carmakers, especially those with no HVs in their current product lineup, are aware of these trends and seem to be focused on securing first mover advantages in bringing this technology to market at affordable prices. Other manufacturers are expected to respond appropriately, generating the conditions for the development of a healthy and sustainable global EV industry.

5

Sustainable Urban Mass Transport

Farheen Qadir

CONTENTS

5.1 Introduction 82
5.2 Rapid Urbanization 83
    5.2.1 Growth of Cities and Urban Sprawl 83
    5.2.2 Environmental, Health, and Social Impact 85
5.3 Transportation and Public Policy...89
    5.3.1 Urban Mass Rapid Transportation..89
    5.3.2 Bus Rapid Transit (BRT) 90
    5.3.3 Public Policy 93
    5.3.4 Institutional Framework 93
    5.3.5 Transport Economics 94
5.4 Case Study: Mexico City—Metrobús 95
    5.4.1 Traffic Jams, Pollution, and Accidents 95
    5.4.2 A Solution: Metrobús 96
    5.4.3 Design 97
    5.4.4 A New Business Model...98
    5.4.5 Operational and Economic Efficiency 100
    5.4.6 Social and Environmental Impact..101
5.5 Lesson Learned and Future Implications...102
    5.5.1 Lesson Learned..102
    5.5.2 Future Implications for the MENA Region—A Closer
           Look at Cairo 103
5.6 Conclusions...105

5.1 INTRODUCTION

According to estimates from the World Bank's *World Development Report*, an increasing share of the world's population has moved from rural to urban areas. Absorbing the 2.4 billion new urban residents expected over the next 30 years will require extensive investments in housing, water, sanitation, transportation, power, and telecommunications (World Bank 2000, 9). The need for these new infrastructure investments comes on top of the backlog that already plagues cities worldwide (World Bank 2000, 132). Moreover, countries should not and cannot wait until they become wealthy to improve urban services. Innovative institutional arrangements including public-private partnerships (PPPs) can yield substantial infrastructure development and allow public services to keep pace with increasing demand.

This chapter focuses on one of the increasingly important problems in fast-growing urban areas worldwide: green urban mass transportation. The chapter begins by providing a brief background on the growth of cities and unfettered urban sprawl, with a special focus on developing economies. What follows is a description of the development of public transport in response to rapid urban population growth with a brief look at common forms of mass transport and their shortcomings. This sets the stage for the introduction of bus rapid transit, or BRT, one viable alternative solution for an increasingly car-centric society. A case study of Mexico City's Metrobús provides a particularly interesting example of an application of BRT in one of today's mega-cities. In Mexico City, one of the largest cities in the world, car use doubled between 2001 and 2009, by which time it was estimated that every day 600 new cars entered the city's streets (EMBARQ 2009b). To address the environmental, health, and social implications of increased traffic and congestion, a unique public-private sector partnership resulted in the successful implementation of an urban mass transit solution, and just as importantly, began to challenge the local notion that mass transit is for the poor. Finally, the chapter postulates on the opportunities for implementation of BRT-type systems in the Middle East and North Africa (MENA) region, such as Cairo, another mega-city in the developing world. Data were obtained both through secondary research (information sourced throughout) and primary research through interviews with parties involved.

5.2 RAPID URBANIZATION

5.2.1 Growth of Cities and Urban Sprawl

According to United Nations projections, virtually all of the world's population growth over the next 30 years will occur in urban areas and by 2025, almost 60% of the world's population will reside in urban centers. Of the urban dwellers of the future, nearly 90% will be living in developing countries. Half a century ago just 41 of the world's 100 largest cities were in developing countries. By 1995 that number had risen to 64, and the proportion keeps rising (World Bank 2000, 9). Figure 5.1 presents historical data and projected trends in worldwide urban and rural population, illustrating this significant shift in world demographic distribution. Table 5.1 is a list of the 50 most populous urban areas in the world, and Table 5.2 the ranking of the 50 fastest growing urban areas. Examination of these two tables provides some interesting insights. For example, while 12 of the 50 cities with the largest population in the world are in OECD countries, only 2 of the 50 cities with largest population growth rates in the world are located in those countries.

Furthermore, urban spatial expansion, or urban sprawl, is a reality in most developing countries. It results mainly from three powerful forces: a growing population, rising income, and falling commuting costs (Brueckner 2000). This urban expansion often occurs without control or planning and exacerbates the mobility challenges that residents of developing cities face, resulting in long commutes, traffic congestion, and air pollution.

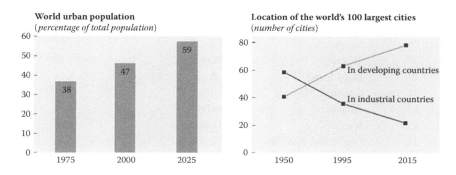

FIGURE 5.1
Urban population is growing, primarily in developing countries. (World Bank, *World Development Report 1999/2000*, New York: Oxford University Press, 2000.)

TABLE 5.1

World's Most Populated Urban Areas in 2006

Rank	City/Urban Area	Country	Population in 2006 (Millions)
1	Tokyo	Japan	35.53
2	Mexico City	Mexico	19.24
3	Mumbai (Bombay)	India	18.84
4	New York	United States	18.65
5	São Paulo	Brazil	18.61
6	Delhi	India	16
7	Calcutta	India	14.57
8	Jakarta	Indonesia	13.67
9	Buenos Aires	Argentina	13.52
10	Dhaka	Bangladesh	13.09
11	Shanghai	China	12.63
12	Los Angeles	United States	12.22
13	Karachi	Pakistan	12.2
14	Lagos	Nigeria	11.7
15	Rio de Janeiro	Brazil	11.62
16	Osaka, Kobe	Japan	11.32
17	Cairo	Egypt	11.29
18	Beijing	China	10.85
19	Moscow	Russia	10.82
20	Metro Manila	Philippines	10.8
21	Istanbul	Turkey	10
22	Paris	France	9.89
23	Seoul	South Korea	9.52
24	Tianjin	China	9.39
25	Chicago	United States	8.8
26	Lima	Peru	8.35
27	Bogotá	Colombia	7.8
28	London	UK	7.61
29	Tehran	Iran	7.42
30	Hong Kong	China	7.28
31	Chennai (Madras)	India	7.04
32	Bangalore	India	6.75
33	Bangkok	Thailand	6.65
34	Dortmund, Bochum	Germany	6.57
35	Lahore	Pakistan	6.57
36	Hyderabad	India	6.34
37	Wuhan	China	6.18

(continued)

TABLE 5.1 (continued)

World's Most Populated Urban Areas in 2006

Rank	City/Urban Area	Country	Population in 2006 (Millions)
38	Baghdad	Iraq	6.06
39	Kinshasa	Congo	5.89
40	Riyadh	Saudi Arabia	5.76
41	Santiago	Chile	5.7
42	Miami	United States	5.48
43	Belo Horizonte	Brazil	5.45
44	Philadelphia	United States	5.36
45	St. Petersburg	Russia	5.35
46	Ahmadabad	India	5.34
47	Madrid	Spain	5.17
48	Toronto	Canada	5.16
49	Ho Chi Minh City	Vietnam	5.1
50	Chongqing	China	5.06

Source: City Mayors, http://www.citymayors.com/statistics/urban_2006_1.html.

The economic success of cities is based on agglomeration economies, which enable businesses to function more efficiently in proximity to a dense network of information, employees, suppliers, and customers. As a result, it is not surprising that cities depend on good transportation systems to thrive. Furthermore, a healthy economy requires a combination of public and private modes of transportation to meet the varied needs of the population. However, the high labor intensity of public transport, combined with a variety of pressures toward more dispersed trip patterns, subjects this basic service to severe cost pressures, which occasionally erupt in cutbacks or unsustainable fiscal drains. Meanwhile, the increase in private car use and traffic congestion continues unabated. The adverse environmental, health, and social effects of congestion limit the sustainability of private modes of transportation in the long term. Thus, an integrated urban plan including sustainable mobility solutions is essential to the survivability of tomorrow's urban centers.

5.2.2 Environmental, Health, and Social Impact

Problems of inadequate infrastructure have economic as well as human costs. As traffic continues to clog the streets of most large cities in developing countries, the costs of traffic congestion grow. Estimated losses

TABLE 5.2

Projected 50 Fastest Growing Urban Areas 2006–2020

Rank	City/Urban Area	Country	Avg. Annual Growth 2006 to 2020, in %
1	Beihai	China	10.58
2	Ghaziabad	India	5.2
3	Sana'a	Yemen	5
4	Surat	India	4.99
5	Kabul	Afghanistan	4.74
6	Bamako	Mali	4.45
7	Lagos	Nigeria	4.44
8	Faridabad	India	4.44
9	Dar es Salaam	Tanzania	4.39
10	Chittagong	Bangladesh	4.29
11	Toluca	Mexico	4.25
12	Lubumbashi	Congo	4.1
13	Kampala	Uganda	4.03
14	Santa Cruz	Bolivia	3.98
15	Luanda	Angola	3.96
16	Nashik	India	3.9
17	Kinshasa	Congo	3.89
18	Nairobi	Kenya	3.87
19	Dhaka	Bangladesh	3.79
20	Antananarivo	Madagascar	3.73
21	Patna	India	3.72
22	Rajkot	India	3.63
23	Conakry	Guinea	3.61
24	Jaipur	India	3.6
25	Maputo	Mozambique	3.54
26	Mogadishu	Somalia	3.52
27	Gujranwala	Pakistan	3.49
28	Delhi	India	3.48
29	Pune (Poona)	India	3.46
30	Las Vegas	United States	3.45
31	Addis Ababa	Ethiopia	3.4
32	Indore	India	3.35
33	Faisalabad	Pakistan	3.32
34	Rawalpindi	Pakistan	3.31
35	Brazzaville	Congo	3.29
36	Peshawar	Pakistan	3.29
37	Khulna	Bangladesh	3.24

(continued)

TABLE 5.2 (continued)

Projected 50 Fastest Growing Urban Areas 2006–2020

Rank	City/Urban Area	Country	Avg. Annual Growth 2006 to 2020, in %
38	Suwon	Republic of Korea	3.23
39	Karachi	Pakistan	3.19
40	Asuncion	Paraguay	3.17
41	Lahore	Pakistan	3.12
42	Asansol	India	3.11
43	Riyadh	Saudi Arabia	3.09
44	Dakar	Senegal	3.06
45	Multan	Pakistan	3.06
46	Valencia	Venezuela	3.05
47	Jakarta	Indonesia	3.03
48	Brasilia	Brazil	2.99
49	Port-au-Prince	Haiti	2.98
50	Palembang	Indonesia	2.94

Source: City Mayors, http://www.citymayors.com/statistics/urban_growth1.html.

from traffic jams in Bangkok range from $272 million to $1 billion a year, depending on how the value of time lost in congested traffic is computed. In Seoul, the aggregate value of lost time from traffic congestion is estimated at $154 million per year (World Bank 2000, 141–142).

Urban air pollution, which has worsened in most large cities in the developing world over the last few decades, imposes a heavy burden on the health of urban populations throughout the developing world. Increased incidence of lung cancer and cardiovascular and respiratory diseases are but some of the long-term health impacts of sustained exposure to outdoor air pollution. Other adverse health effects include increased incidence of chronic bronchitis and acute respiratory illness, exacerbation of asthma and coronary disease, and impairment of lung function. For children in large cities of developing countries, breathing the air may be as harmful as smoking two packs of cigarettes a day. In Delhi, the incidence of bronchial asthma in the 5 to 16 age group, of which air pollution is one of the major causes, is 10% to 12% (World Bank 2000, 141). Nine of the 10 cities with the highest counts of total suspended particulates (TSPs) in the world are located in China (World Bank 2000, 141–142). Industrial and industrializing cities such as Jiaozhou, Lanzhou, Taiyuan, Urumqi, Wanxian, and Yichang all have mean annual concentrations of TSPs exceeding 500 micrograms per cubic meter. The World Health Organization (WHO) puts acceptable levels at less than 100 micrograms per cubic meter. If China maintained its former business-as-usual response to air pollution, the health costs of urban residents' exposure to TSPs would rise from $32 billion in 1995 to nearly $98 billion in 2020 (World Bank 2000, 142). Fortunately, recently there has been a concerted effort on the part of the Chinese central government to address the pollution problem.

Transport is a known cause of many air pollutants. The following is a description of classic transport-related air pollutants:

- Lead from the combustion of leaded gasoline is a well-known toxin. High levels of lead in the bloodstream may increase incidence of miscarriages, impair renal function, and increase blood pressure. Most significantly, it may retard the intellectual development of children and adversely affect their behavior (Gwilliam, Kojima, and Johnson 2004, 1).
- Total suspended particles are not a single pollutant, but rather a mixture of many subclasses of pollutants that occur in both solid and liquid forms. Each subclass contains many different chemical compounds. Emerging scientific evidence points to increasing damage with decreasing particle diameter. Particles larger than about 10 μm are deposited almost exclusively in the nose and throat, whereas particles smaller than 1 μm are able to reach the lower regions of the lungs.
- Ozone (O_3) has been associated with transient effects on the human respiratory system, especially decreased pulmonary function in individuals taking light-to-heavy exercise. Several recent studies have linked ozone to premature mortality. Ozone also reduces visibility, damages vegetation, and contributes to photochemical smog. Oxides of nitrogen (NO_x) and volatile organic compounds (VOCs) are the two main precursors of ozone. NO_x is emitted by gasoline- and diesel-powered vehicles, while VOCs are emitted in most significant quantities by gasoline-fueled vehicles.
- Carbon monoxide (CO), to the expansion of which gasoline-fueled vehicles are the largest contributors, inhibits blood capacity to carry oxygen to organs and tissues. People with chronic heart disease may experience chest pains when CO levels are high. At very high levels, CO impairs vision and manual dexterity, and can cause death.
- Sulfur dioxide (SO_2), which is emitted in direct proportion to the amount of sulfur in fuel, causes changes in lung function in persons with asthma and exacerbates respiratory symptoms in sensitive

individuals. Through a series of chemical reactions, SO_2 can be transformed to sulfuric acid, which contributes to acid rain.

- Nitrogen dioxide (NO_2) also causes changes in lung function in asthmatics. Like SO_2, NO_2 can react to form nitric acid and thereby contribute to acid rain.
- Other air toxin emissions of primary concern in vehicle exhaust include benzene and poly-aromatic hydrocarbons (PAHs), both well-known carcinogens.

According to the United Nations, the greatest payoff toward reducing air pollution is expected to come from channeling urban growth along transit routes to create more efficient transportation corridors (World Bank 2000, 10).

5.3 TRANSPORTATION AND PUBLIC POLICY

5.3.1 Urban Mass Rapid Transportation

Mass rapid transit includes bus-based and rail-based modes operating on a right-of-way basis with substantial exclusivity, that is, where only limited sections or intersections are also used by other forms of traffic. Electrically propelled transport modes are the least polluting forms of mass transit. These systems, including subway, light rail, and rail give a perception of permanence and quality. Unfortunately, electric rail-based systems are expensive. Recent investments in underground rail or subway have cost between $40 and $100 per km, which is beyond the resources of most developing cities. Moreover, due to the substantial investment in infrastructure required, such as laying tracks and building stations, electric rail projects require 5+ years to complete. Given the pressure politicians feel to show results within a political term, there is often a strong hesitation on their part toward commitment to projects with long lead times (Gwilliam et al. 2004, 63).

Many studies have compared rail transit to other modes of transport. In most cases, these studies have found that buses can supply the same service as rail at a fraction of the cost. Yet many cities still choose to invest in rail due to pressures from private interest groups, the desire to enhance a city's image, the advantages of rail in terms of comfort and convenience, and/or support from federal grants. In contrast, buses typically require very little investment in infrastructure and very short lead times to begin operation. Unfortunately, buses are often synonymous with uncomfortable, dirty, polluting, and inefficient transport. Furthermore, in developing countries, buses are most commonly operated via an owner-operator model whereby many individual operators provide bus service in an uncoordinated, inefficient, and certainly suboptimal manner. However, if designed and managed correctly, buses can be a promising alternative to rail (Gwilliam et al. 2004, 64). Recent innovations in the design of bus transit have begun to address some of the issues of comfort and convenience and go a long way toward giving bus transit some of the characteristics of rail transit. These innovations collectively comprise a system known as bus rapid transit, or BRT.

5.3.2 Bus Rapid Transit (BRT)

BRT is a cost-effective, safe, clean, mass transit alternative. It is essentially a transport system that combines characteristics of bus transport with those of rail transport. Similar to buses, BRT requires minimal infrastructure investment; however, like trains and subways, it provides efficient service.

The main attributes that differentiate BRT from conventional bus service are the following:

- Dedicated lanes. According to the World Bank, dedicated lanes or totally segregated bus-ways are essential to counteract the problems of mixed traffic, specifically, the reduction in the average speed of traffic when multiple modes of transport share one roadway (Gwilliam et al. 2004, 59).
- Rigorous planning, operational controls, and modal integration. It has been shown that good transport planning and service integration are essential to the success of BRT (Gwilliam et al. 2004, 63).
- Traffic management through timed signaling. It has been shown that with good traffic management to minimize car delays, segregated bus systems can produce both efficiency gains and environmental benefits (Gwilliamet al. 2004, 60).
- Rapid boarding and exiting, and prepayment of fares.
- High-capacity modern vehicles sometimes with clean technology. No specific technology defines the buses in the BRT system, nor does the system require clean technology to have a positive impact on the environment. The environmental benefits arise from (1) providing

high-quality mobility service such that car riders are motivated to leave their cars at home and take the BRT system; and (2) requiring that buses be properly maintained using preventive maintenance measures. Any investment in clean technology for buses provides an additional environmental benefit of the system.

BRT is not new. In 1974, visionary city planners under the guidance of Curitiba, Brazil's then-mayor Jaime Lerner, an architect and urban planner, developed the first BRT system, a surface bus system with the advantages of a subway. Curitiba, the seventh largest city in Brazil, with a metropolitan population of over 3 million in 2009, is one of the very few examples of a city where a transport system was developed alongside and simultaneously to urban growth rather than in reaction to it. By channeling urban growth along mass transit routes, the city has reduced the use of private cars—despite having the second highest rate of per capita car ownership in Brazil.

A convincing example of how integrated public planning can improve accessibility at relatively low cost, Curitiba's BRT carries 2.2 million passengers per day and is used regularly by more than one third of the city's population (EMBARQ 2006a, 18). As a result, Curitiba's gasoline use per capita is 25% lower than that of eight comparable Brazilian cities, and the city has one of Brazil's lowest rates of ambient air pollution (World Bank 1998). Figure 5.2 compares the length of dedicated lanes for public transport per inhabitant for various cities in 2006, graphically illustrating the privileged position Curitiba finds itself in.

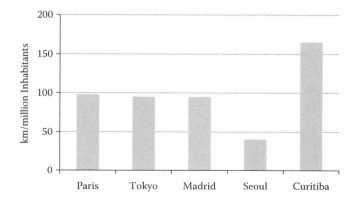

FIGURE 5.2
Length of public transport dedicated lanes. (World Bank, September 6, 2006.)

TABLE 5.3

Cost per Kilometer for Various Rapid Mass Transportation Systems

	Cost per km (Millions of USD/km)
Bus rapid transit	7
Light rail	15
Subway	40

Source: Hidalgo, D., High Level BRT: *An Option to Consider Even at Very High Demand Levels,* Paper presented at the 5th International Bus Conference, Bogotá, Colombia, February 14, 2006.

From a cost standpoint, BRT is highly competitive with the alternatives. It can satisfy a demand of between 20,000 and 40,000 passengers/hour/ direction, comparable to metro systems, but with an initial set-up cost of $5 million to $10 million per kilometer (km) versus $30 to $160 million for a metro system (Hidalgo 2006). Table 5.3 shows a comparison of cost per km for bus, light rail, and subway rapid transit. It is clear that there is a significant cost advantage to implementing BRT over light rail or subway.

BRT projects are preferred by municipal districts because they require less capital and a shorter lead time to become fully operational. A BRT project can be conceptualized, designed, and implemented within the term of one administration, which is a big advantage over subway systems which can take up to 10 years to implement. BRT systems have become the urban mobility system of choice in 23 countries on five continents. Currently there are more than 70 systems either operating or under construction. Several cities in the United States have recently invested in upgrading existing bus systems to BRT systems. Los Angeles and Vancouver have both seen early improvements in transit time and average speed on the corridor. The Australian cities of Sydney, Adelaide, and Brisbane have all reported travel time savings of 37% or more, substantial patronage growth, and a positive effect on property values adjacent to the BRT corridors (Graham 2006, 7).

Moreover, BRT is often not merely a bus service. It provides an excellent opportunity to improve the economic and urban environment through investment in public spaces as well. By developing plazas, parks, sidewalks, and bike paths not only is there an urban renewal aspect to the project with social benefits for the population, but also there is a stimulus toward increased economic activity resulting in economic renewal. The overall result of an integrated BRT implementation is a marked improvement in the quality of life for all involved.

5.3.3 Public Policy

Automobile use increases as incomes rise and cities grow spatially, weakening mass transit systems. The major problems of urban transportation relate to traffic congestion, pollution from emissions, and limited mobility. The appropriate policies for addressing these issues require urban governments to optimize land use, to manage traffic and demand for transportation, to formulate environmental policies and measures to mitigate congestion, to improve fuel efficiency, and to set up vehicle emissions control and inspection systems (Stiglitz 1997).

Perhaps the greatest payoff is from integrated land use and transportation planning: new roads open doors to land development, and compact urban centers increase the possibilities for mass transit. However, coordinating transportation and land use policies remains politically difficult in many developing countries. A start could be made in urban areas where motor vehicle ownership is still low, land remains available, and land use patterns are still evolving. Even cities with high rates of automobile ownership can develop efficient transportation alternatives that accommodate the needs of all social groups. Many cities have combined innovations in mass transit with effective planning and controls for automobile use: Copenhagen; Curitiba; Freiburg, Germany; Hong Kong; Perth, Australia; Portland (Oregon), United States; Singapore; Surabaya, Indonesia; Toronto; and Zurich (World Bank 2000, 150).

5.3.4 Institutional Framework

An institutional framework is required to ensure success of the BRT system. This framework consists of a clear definition of roles and responsibilities for each of the various public and private entities involved. It also requires rules and regulations upon which working relationships and negotiations are based.

A key to success is the participation of existing bus owners and operators in the new system. A BRT implementation has the ability to improve the organizational structure of existing bus companies, to clarify relationships between service providers and authorities, and to create a transit system that is sustainable in the long term. However, this requires open channels of communication, and transparency throughout the project, to build a trusting relationship and open dialogue between bus operators and the city.

BRT offers the city an alternative to patchworks of poorly regulated and inefficient bus systems. The BRT model requires individual bus owners and drivers to join together under a more formal business model. This brings about greater organization and coordination, which translates into more regular income and formal employment status for the bus drivers, that is, access to government benefits and better, more reliable service for the passenger. Improving the system also requires a new economic model that breaks the link between passenger and income. BRT addresses this issue through a pay per km system where deviations from the service schedule result in heavy penalties. In some implementations of BRT, the newly organized bus owners compete against other regional, national, and even international companies for the right to provide bus service within the BRT framework.

5.3.5 Transport Economics

Two revenue streams exist in public transport systems: the government and the passenger. Governments currently pay for the building of roads which, at least in developing countries, transport the wealthiest residents. In turn, they should invest in the infrastructure for public transport systems to provide mobility options for the rest of the residents as well. Passenger fares then are set to cover operational costs on a day-to-day basis so as to ensure long-run sustainability.

In the absence of road pricing, many cities subsidize urban rail transport to compensate for its high cost. Across the 20 largest transit systems in the United States, the subsidy as measured by the difference between operating cost and passenger fare revenues, ranged from 29 to 89% for rail and 57 to 89% for bus (Kerin 1992). Two rationales are often provided for this. First, the marginal cost in a mass transport system is usually below the average cost per user of the system given economies of scale, so it makes sense that users should be incentivized to use these systems over private automobiles through below-cost pricing. The economies of scale arise from fixed costs, such as track and station maintenance, and from the fact that higher passenger density allows for vehicle operation at or near full capacity, therefore permitting transit providers to more easily amortize costs.

The second rationale is that lower transit fares are expected to discourage private automobile use in favor of public transport use. As a result, external costs of traffic congestion, such as air pollution and traffic accidents, are reduced. This assumes that these costs are not taken into account in congestion pricing and thus peak-hour private automobile use is under-priced, creating an inefficient amount of congestion on roads. That is, most cities do not have congestion pricing or tolls during peak hours to control road use.

Opponents of subsidies point to several disadvantages. First, it is hard to generate a large enough differential between cost of driving and cost of public transport through subsidies alone. Second, subsidies are a drain to urban fiscal budgets, whereas road tolls are a revenue source. This is particularly important for cities in developing countries, which may have competing priorities for limited funding. Third, subsidies to public transport encourage urban sprawl and thus may have the perverse second-order effect of stimulating the long-term dependency on private transport (Gwilliam et al. 2004, 64).

When BRT systems are placed in high-demand corridors and are integrated with the city's existing transportation networks, operating revenues meet or exceed operating costs, making subsidies unnecessary and raising the attractiveness of these systems. One example of this is Mexico City's Metrobús.

5.4 CASE STUDY: MEXICO CITY—METROBÚS

5.4.1 Traffic Jams, Pollution, and Accidents

Home to 18 million people and 6 million cars, Mexico City is one of the largest cities in the world and undoubtedly one of the most congested (EMBARQ 2009b). "Travel in Mexico City is a universal affliction that hurts the individual, the community and the environment" (EMBARQ 2006b, 30). Commuters spend an average of two and a half hours in traffic per day, according to the Mexican National Institute of Ecology (Sustainable Mobility 2006b, 30–31).

Private cars make up 90% of the vehicular activity in the city and generate 50% of the pollution even though they transport only 20% of all travelers. Internal combustion engines are responsible for 80% of all the air pollution and generate 30% of the region's carbon dioxide emissions, according to data from the Mexico City Secretary of Environment. Mexico City, in part by virtue of its altitude and geography, has historically suffered from high ambient concentrations of ozone, with levels exceeding the acceptable limits on 225 days of the year (EMBARQ 2006c, 34). Furthermore, the government has been imposing increasingly stringent emission standards on its 3 million gasoline-fueled vehicles, aided by a strictly enforced emissions inspection program. Additionally, auto accidents result in 2,500 deaths a year in the Mexico City metropolitan area.

Although the increased use of private cars is associated with increased congestion, pollution, and adverse health effects, Mexico City's existing public transportation options have their own problems. Bus riders must suffer noisy, polluting buses whose drivers take erratic routes through the city streets. Subway riders must endure suffocating crowds, deteriorating infrastructure, and frequent service interruptions (SETRAVI 2009).

In 2001 Mario Molina, recipient of the Nobel Prize in Chemistry and founder of the Mario Molina Institute for Strategic Studies on Energy and the Environment, proposed that the city focus on public transport in order to address its air pollution problem. Shortly thereafter, a private-public partnership was formed to improve transportation and environmental conditions in Mexico. The partners in this movement include the Mexican National Institute of Ecology, an independent government agency; EMBARQ, the World Resources Institute for Sustainable Transport; the Center for Sustainable Transport (CTS) in Mexico; and the Mario Molina Center for Strategic Studies on Energy and Environment. Together, they developed a four-pronged strategy to address the downward spiral that has characterized traffic and pollution in Mexico:

1. Reduce automobile use by developing alternative transportation options.
2. Promote non-motorized transport solutions such as walking and bicycling.
3. Promote a high-capacity, efficient, clean, safe, comfortable, and accessible public transport system.
4. Promote more efficient and less polluting vehicle and fuel technologies.

5.4.2 A Solution: Metrobús

Metrobús is a poignant example of a successful private-public partnership. CTS, a Mexican non-governmental organization (NGO) affiliated with EMBARQ of the World Resources Institute, partnered with the federal government of Mexico and with Mexico City's local government to implement the first BRT corridor in Mexico City. Construction of the Metrobús began in January 2005 with service beginning only 6 months later, in June 2005. Currently approximately 450,000 passengers per day travel using BRT (EMBARQ 2009c).

Prior to Mexico City's Metrobús project, a BRT line was designed in Leon. Optibus, as it is called, started operation in 2002. After Metrobús, in March 2009, the Macrobús system in Guadalajara began operation. It currently transports 90,000 passengers per day on a network of 16 km (EMBARQ 2009a). Additionally, two projects are currently ongoing: one in the State of Mexico and another in Chihuahua. Finally, other Mexican cities are currently competing for federal funding for public transportation projects. Given the success of the BRT lines in Leon, Mexico City, and Guadalajara, it is not surprising that the overwhelming majority of these project proposals are for BRT systems.

5.4.3 Design

The inaugural Metrobús corridor ran along a 20-km section of Avenida Insurgentes, a central north-south transport artery and a major commercial thoroughfare for the city. Two other lines have since been added, one that effectively extends the first line southward an extra 9 km and another one that runs east-west in the southern part of the city. In addition, a fourth line was under construction in late 2009. Mexico City mayor Marcelo Ebrard has committed to creating a citywide network totaling 10 lines spanning over 120 km by 2012. CTS continues to work with the municipal and national governments in Mexico on not only transport projects but also air quality, communications, mobility, and urban planning initiatives across Mexico.

The buses in the Mexican fleet are single- and double-articulated diesel buses. An articulated bus is comprised of two rigid parts linked by a pivoting joint. This accordion-like design allows for a longer legal bus length and thus a greater passenger capacity than standard buses. Single-articulated buses have a capacity of about 160 passengers, whereas double-articulated buses have a capacity of approximately 200 passengers. The Mexican fleet of buses runs on clean burning, ultra low sulfur diesel that meets Euro III emissions standards. Under European emissions standards, emissions of nitrogen oxides (NO_x), total hydrocarbon (THC), non-methane hydrocarbons (NMHCs), carbon monoxide (CO), and particulate matter (PM) are

FIGURE 5.3
Single-articulated buses. (Metge, H. and Jehanno, A., *A Panorama of Urban Mobility Strategies in Developing Countries*, Washington, DC: World Bank, 2006.)

regulated (see Figure 5.3 for pictures of Mexico City's Metrobús, which has state-of-the-art design).

They travel in a dedicated, protected corridor, where unauthorized vehicles are subject to stiff fines. Additionally, traffic lights are preferentially timed to minimize delays to the Metrobús. Stations are raised central platforms at an average distance of 450 meters to minimize boarding times. With the addition of the third Metrobús corridor, the Metrobús offered connections with all Metro lines. Passengers purchase a rechargeable prepaid smartcard and use it to access platforms via turnstiles, thereby eliminating delays arising from payment at boarding. The buses run 24 hours a day with an interval of only two minutes during rush hour.

5.4.4 A New Business Model

In Mexico, as in many Latin American countries, transport services are provided by individuals who receive concessions from the state and may own one or more vehicles. This system is characterized by a large number of owners and very little organization or coordination. Owners typically rent out their bus or buses to drivers, who often are paid by the number of passengers they transport. The result is a fierce competition among bus drivers who drive aggressively, stop as needed, invade dedicated lanes, and create traffic jams in search of the additional passenger. They generally require government subsidies and minimize preventive vehicle maintenance, which is not a sustainable *modus operandi*.

Prior to the implementation of Metrobús, 70% of the bus transport demand along Avenida Insurgentes was served by privately owned buses through concessions as described above with the remainder (30%) served by government-owned buses. The government encouraged and to some extent forced these individual owner-operators to come together and form a transport company. This required that bus owners sell their old buses and that the transport company buy new articulated buses according to BRT requirements. The bus drivers would then be official employees of the transport company. The transport company took over 70% of the BRT bus demand as before. The other 30% was given to a public operator meeting the same requirements as the private transport company. In this way, the government maintained the same supply dynamic as before.

In Mexico City, negotiations with the current operators were direct—the project required their commitment and would only work with their full support. This approach was required because of the size and political influence of the existing owner-operators. In Mexico City, there are 30 million trips daily of which 70% are done using public transport (Y.K. Voukas, interview, August 26, 2009). Of these trips, 70% are served by private companies. That is, over 20 million trips were controlled by private operators of public transport. Prior to the Metrobús project, 60,000 individual private operators served this demand. These 60,000 operators, their families, and those who relied indirectly on this economic activity all had a vested interest in the owner-operator model. It would have been political suicide to marginalize this group and move for an open bid. Thus, direct negotiations through a non-bid process were favored. This meant convincing the operators that they would be better off in the long run under the new model through improved employment status, greater government benefits, and better working conditions.

In contrast, in the city of Guadalajara, the existing transport companies were already formally organized around common business principles, so the requirement of Macrobús to form transport companies meant little change for them. Additionally, the government leaders were stronger than the unions and had greater influence than in Mexico City. An open bid was held where the local transport company had to bid against international operators to win the right to continue operation of the corridor. However, the local company was told it would be favored in this open bid because it was able to meet certain standards. This process allowed for competition and thus inherently resulted in greater efficiency and lower costs.

5.4.5 Operational and Economic Efficiency

Prior to the Metrobús, the daily demand along the Avenida Insurgentes corridor of 220,000 passengers was being served primarily by 380 old, poorly maintained, polluting buses. With the implementation of BRT, the same demand was served with 80 articulated buses with modern fuel technology. Within a few months, the demand increased to 260,000 passengers per day, and nine additional buses were added. These additional passengers were primarily those who used to travel on parallel services but, having seen the reliability of the Metrobús, preferred to travel the length of Insurgentes using the Metrobús and then move to a connecting service to get to their final destination. Additionally, some of the demand was that of Metro passengers who preferred the Metrobús even though the average travel time was longer because there was much less variability. The Metro is notorious for having mechanical problems—a ride that took 40 minutes one day could take 1 hour and 40 minutes the next. In contrast a 1-hour ride on the Metrobús might take 5 to 10 minutes less or more from one day to the next. In mid-2009, as supply continued to outstrip demand, and with the total number of buses possible in the corridor at its maximum, single-articulated buses were being replaced by bi-articulated buses with higher capacity to address the increase in demand.

In the first year of operation, a customer survey of Metrobús riders showed that approximately 10% of Metrobús passengers used to drive the same route (Y.K. Voukas, interview, August 26, 2009). So, given fewer cars on the road, fewer buses on the road, and a more orderly driving pattern of buses due to predetermined stops and a dedicated travel lane, driving conditions improved considerably. Average car speed in the Avenida Insurgentes corridor increased from 12 to 16 km/hr. Consequently, this improvement made roads more attractive to drivers, and some of those who had originally switched from driving to taking the Metrobús switched back to driving. The resulting percentage of Metro riders that previously drove cars dropped to 4% and finally reached an equilibrium number of 6%.

Efficiency in transportation is measured by the IPK index. IPK is the number of passengers per day divided by total km traveled per day. At 300,000 passengers per day and 30,000 km traveled per day, the first corridor has an IPK of 10. Most other BRTs operate around IPK values of 4 to 8. A high IPK means buses are running full, which means high revenues, which in turn implies the operational and financial success of the corridor.

TABLE 5.4

Public Transport Subsidies

Mode	Passenger Fare (Pesos)	Government Subsidy per Passenger (Pesos)
Metro	2	3.5
Bus	2	5
Trolley bus	2	11
Metrobús	5	0

Source: Interview with Yorgos K. Voukas, August 26, 2009.

However, there is also the implication that quality suffers since buses are running at or near capacity all the time.

As mentioned earlier, most public transport systems around the world are subsidized. Operational costs are significantly higher than operational revenues, so the government covers the difference in order to continue providing a public service. The case is no different in Mexico City. All public transportation systems in Mexico City require subsidies to continue operation. All except for the Metrobús, that is. Table 5.4 shows a comparison of fares and subsidies for various forms of transport in Mexico City.

The Metrobús can justify a higher fare than other modes because of the significantly higher level of service it provides. Metrobús is clean, safe, and reliable, which allows it to stand out from other modes. The revenues brought in by the passenger fares alone cover all operating costs of the Metrobús, including bus drivers, maintenance, and regulating authority fees.

5.4.6 Social and Environmental Impact

The benefits of the BRT system are multifold. Individual passengers benefit from shorter travel times, improved safety and comfort, and increased reliability. Travel time from one end of the Avenida Insurgentes corridor to the other has been reduced by 50%. Given average travel distances for riders, this has meant a reduction in travel time of 15 minutes per trip. Furthermore, because the fleet of buses is new and the operators are loosely evaluated on the quality of service they provide, the Metrobús scheme provides a more comfortable ride and a higher quality of service for riders. Finally, Metrobús provides greater security to the passengers when compared to Metro stations. There is far better lighting on and around platforms, commercial activity is forbidden, and the entire travel experience takes place above ground where there is higher visibility and transparency.

Concessionaries or individual bus owners become experienced small business owners and have the potential to run more profitable businesses. Drivers, as a result of gaining formal employment status, become eligible for government benefits, work a shorter work week, and enjoy greater respect in the driver-passenger relationship. Additionally, they get better financial and legal protections, job security, and better working conditions. Society benefits from improved air quality, a better urban image, and improved productivity as previously lost hours are regained.

Benefits for the government are fast infrastructure construction, minimal investment compared with alternative public transport options, improved road safety, and increased control over public transport. Furthermore, Metrobús is the first public transport system in the world to receive carbon credits through the carbon credit market. The government of Mexico City has received 150,000 Euros from Spain's Carbon Fund as retribution for the reduction in carbon dioxide emissions through Metrobús. This compensation forms part of an agreement signed with the World Bank in 2005 in which Mexico City assumed the commitment to reduce carbon dioxide emissions by 34,000 tons per year. Currently, Metrobús reduces an estimated 80,000 tons of CO_2 emissions per year (Y.K. Voukas, interview, August 26, 2009). In only 4 years, Metrobús has already begun to make a positive impact on the environment. It is estimated to reduce nitrogen oxide pollution by 690 tons per year, hydrocarbon pollution by almost 150 tons per year, besides reducing an estimated 2.8 tons of fine particulate matter (EMBARQ 2009c).

5.5 LESSON LEARNED AND FUTURE IMPLICATIONS

5.5.1 Lesson Learned

The success of the Metrobús initiative was not coincidental. One important lesson learned is that a strong public-private partnership can provide the impetus for change and encourage multi-lateral cooperation in urban infrastructure development.

The Metrobús initiative is the result of successful collaboration among several public and private institutions. EMBARQ (The World Resources Institute Center for Sustainable Transport) worked on the project with three Mexico City government agencies—the Secretary of the Environment,

Secretary of Transportation, and Secretary of Urban Development—and the Interdisciplinary Center for Biodiversity and Environment (CeIBA), a prominent Mexican NGO. Together, these groups established the Center for Sustainable Transport in Mexico (CTS-México), a not-for-profit organization that has provided ongoing technical assistance to the Metrobús system from its inception through its expansion. Furthermore, significant financial support was provided by the Shell Foundation, Caterpillar Foundation, the William and Flora Hewlett Foundation, and the World Bank.

In the case of Mexico City, the private sector participation helped overcome the political challenges that often impede sustainable transportation. Disparate groups came together and worked toward cooperation and compromise rather than competition. "Each partner contributed to the success of Metrobús in different ways," as Nancy Kete of EMBARQ described. "By coming in from the outside, EMBARQ brought new ideas, international prestige, global best practices, and helped arrange financing. CTS-México has provided an experienced voice to support, assure and guide the city on a day-to-day basis. In the end, the public will hold the mayor and his senior staff accountable for the quality of transport in the city" (Smith 2009).

In November 2009, the Mexico City Metrobús Project and the partnering institutions that supported this initiative received the 2009 Roy Family Award for Environmental Partnership from the John F. Kennedy School of Government at Harvard University. This award recognizes outstanding public-private partnerships that enhance environmental quality through the use of novel and creative approaches.

5.5.2 Future Implications for the MENA Region—A Closer Look at Cairo

In the MENA region almost 60% of the population lives in cities (World Bank Group 2009). Already, 8 of the region's cities have more than 3 million citizens. The largest, Cairo, has over 11 million. Furthermore, 9 cities ranked in the top 100 fastest growing cities in the world with 2.4 to 5% annual population growth are in the region. With the transformation of the economy and deep-seated social changes currently taking place, most cities have experienced rapid growth in urban transport demand and in motorization. As a result, many of the region's large urban areas, where the bulk of GDP is produced, face increasingly difficult transport problems with a high degree of traffic congestion, reduced mobility, and

FIGURE 5.4
Egypt's existing rail infrastructure. (Metge, H. and Jehanno, A., *A Panorama of Urban Mobility Strategies in Developing Countries,* Washington, DC: World Bank, 2006.)

deteriorating air quality. The problems of congestion, limited mobility, and air pollution reduce social and economic opportunities, lower the quality of life, and worsen a city's competitiveness and economic growth potential.

Cairo, the 17th largest city in the world, has undergone polycentric urban development with major urban activities located not just in the historic center but also in satellite cities, according to the city's master plan. Even though the city has been investing in mass transport, demand is outstripping supply. Two metro rail lines currently exist, and one is under construction. Although there is significant existing unused rail infrastructure, much of it is not in good shape, as exemplified in Figure 5.4, and, due to overdevelopment and high population density in the city center, there is limited space on the ground level to build additional transportation infrastructure. As a result, an appropriate strategy for Cairo might be to invest in the existing rail infrastructure to provide light rail transport within the city and to develop a BRT system to provide connecting transport between satellite cities and the center (Metge and Jehanno 2009).

5.6 CONCLUSIONS

The scale of problems in urban mass transportation, in terms of both mobility and external impact, is largest in mega-cities in developing countries with high growth rates such as Mexico City, Cairo, Bangkok, and Lagos (although they are also present in medium-sized cities). Furthermore, the problems are extremely damaging to the economies of those mega-cities. In general, the situation is the worst in places where population growth rate is higher than the rate of economic growth, notably in all of Sub-Saharan Africa and many Latin American countries. In contrast, some cities in the developing world have conceived and implemented successful urban transport programs that are the envy of the world. Most are in Latin America and are now being emulated on other continents. Curitiba was a pioneer in launching bus-based rapid transit in the mid-1970s, and in making the unique link between land use and transport infrastructure development. More recently, Bogotá carried out a successful shift from a low-quality public transport system based on informal transport operators to Transmilenio, an advanced version of bus-based rapid transit integrated with a high-quality, trunk-and-feeder system using street buses.

Currently BRT systems are in operation across Asia, Africa, Latin America, North America, Europe, and Oceania, and new systems are constantly being proposed and implemented across the world. Some of the largest investments in urban mass transport in the next few years will come from Asia, specifically India and China, as they are looking at BRT implementations in several cities on a large scale to keep up with current unprecedented rates of urbanization and motorization. How they will evolve is of global significance (Mitric 2008).

Improvements in essential urban services, such as sustainable mass transport, offer hope and direction for the future. The challenge is to share the successful experiences and bring noteworthy achievements to every city. Success stories, such as Mexico City's Metrobús, also reaffirm the importance of cities developing appropriate institutions and the advisability of reaping the most benefits from the private sector, community-based organizations, and NGOs.

A thorough look at sustainable urban mass transport cannot ignore the role of efficient comprehensive urban planning to guide investments and to identify appropriate locations for jobs, residences, and transportation. The process can help cities avoid the worst outcomes of unplanned growth. Moreover, an overall strategic plan needs to be followed by coherent decentralized implementation that creates a substantial role for the private sector. This type of careful planning and implementation is particularly important in developing mega-cities, some of which are larger than countries.

The challenge of providing sustainable urban mass transport to tomorrow's mega-cities is enormous and complex, but not impossible. Through increased private-public partnerships, strategic and coordinated urban planning, and appropriate use of mass urban transport modes such as BRT, less congested roads and cleaner air need not be pipe dreams for the masses living in today's developing nation urban areas.

6

The Promise of Sugarcane Ethanol as a Cleaner Combustion Engine Fuel

Jeffrey B. Bryant and María Teresa Burbano

CONTENTS

6.1 Introduction 107
6.2 General Considerations on Ethanol 110
    6.2.1 The Technology of Ethanol Fuel 110
    6.2.2 Four Important Ethanol Fuel Characteristics 111
6.3 Major Producers and Key Players 115
6.4 The Economic and Public Policy Issues of Sugar-Based Ethanol ... 117
6.5 The Role Model: Brazilian Ethanol 119
6.6 A New Ethanol Producer: Colombia 124
6.7 Political Barriers and Market Failure: The United States 125
6.8 Ethanol and Income Equality: Mexico 128
6.9 A Potential Sugar-Based Ethanol Entrant: Cuba 130
6.10 Conclusion 133

6.1 INTRODUCTION

Ethanol is a relatively clean-burning, carbohydrate-based fuel that is being touted as a replacement for nonrenewable and pollution-intensive fossil fuels. It is one of eight emerging, large-scale clean energy sectors critical to the future, as identified by the World Economic Forum (WEF 2009). Perhaps the best-known variant is corn-based ethanol, which is popular in the United States. Corn-based ethanol, however, has become shrouded in controversy, and for good reason, as many have attacked its high production costs and large energy input, which detract from its potential benefits.

These factors have diminished ethanol's ability to take hold in the United States and in other industrialized nations.

The experience of a major emerging economy, Brazil, has been radically different. Its sugarcane-based ethanol fuel has been much more successful than the corn-based variant in achieving both broad-scale consumption in its domestic market and substantial exports abroad. As David Luhnow and Geraldo Samor (2006) of the *Wall Street Journal* argued, "After nearly three decades of work, Brazil has succeeded where much of the indus-trialized world has failed: It has developed a cost-effective alternative to gasoline. Along with new offshore oil discoveries, that's a big reason Brazil expects to become energy independent this year."

Three characteristics of the Brazilian experience with sugarcane ethanol stand out. First, as described below, the sugarcane-based ethanol that is pro-duced and commercialized in Brazil is highly efficient both energetically, with up to six times the energy transfer rate of corn ethanol, and economi-cally, with production costs that are the lowest in the world. Second, it has reached large-scale production. Third, this is an example of an emerging-market success story. Brazil, acting as a first mover, was able to accomplish something that the advanced industrialized world could not.

In 2007, ethanol represented 16.7% of the total energy consumption in a country of nearly 200 million people (Empresa de Pesquisa Energética 2008). The success of flex-fuel vehicles, together with the mandatory E25 blend, allowed ethanol fuel consumption to achieve 50% market share of Brazil's automotive fuel market in the first half of 2008, as Brazil con-sumed 2.4 billion gallons of gasoline and 2.38 billion gallons of ethanol (Rapoza 2008). As of April 2009, the fleet of flex-fuel vehicles running on ethanol reached 7.5 million, representing around 25% of all registered light vehicles (ANFAVEA 2009).

Global reaction to this success story has recently been gaining momen-tum. Countries with large fuel bills, such as India and China, are following Brazil's model closely as a strategy for reducing their dependence on for-eign oil. In addition, Japan and Sweden are importing ethanol from Brazil to help fulfill their environmental obligations under the Kyoto Protocol. The United States, until recently, has taken only small steps toward etha-nol use and has relied to a greater extent on the expensive, less efficient, corn-based ethanol. The administration has made climate change a key priority. President Obama seems to recognize the potential benefits that may be had from adopting some of Brazil's strategies. In fact, at the 2009 Summit of the Americas, held in Port of Spain, Trinidad and Tobago, the

President expressed his admiration for Brazil's accomplishments and his desire to make green technologies, such as ethanol, a key part of regional economic recovery:

> Today, I'm proposing the creation of a new Energy and Climate Partnership of the Americas that can forge progress to a more secure and sustainable future. It's a partnership that will harness the vision and determination of countries like Mexico and Brazil that have already done outstanding work in this area to promote renewable energy and reduce greenhouse gas emissions. Each country will bring its own unique resources and needs, so we will ensure that each country can maximize its strengths as we promote efficiency and improve our infrastructure, share technologies, support investments in renewable sources of energy. And in doing so, we can create the jobs of the future, lower greenhouse gas emissions, and make this hemisphere a model for cooperation. (Obama 2009)

Indeed, Brazil has demonstrated not only that sugarcane ethanol can be produced massively, but also that it has a favorable carbon footprint and contributes to exports, employment, and R&D. Given this broad success, it seems only natural to investigate the possibility of transplanting this system to countries in the Americas that share some of Brazil's characteristics. Possible benefits of the ethanol system that may motivate this type of transition include freedom from imported oil, lower energy costs, and reduced carbon emissions, to name a few.

A multidimensional and cross-disciplinary analysis is required when evaluating the feasibility of a transition to ethanol. Economically, one must consider the market forces and relative prices that provide incentives to actors and potential entrants. Political considerations are also important. Energy transitions are rarely possible without the support of agricultural lobbies and government support through subsidies and tax incentives. In addition, movements to reduce foreign oil imports are usually motivated by politics rather than economics, although this was not the case in Brazil. Finally, sociopolitical issues, such as how the positive environmental effects of ethanol resonate with the general public, also play a role.

As previously mentioned, there has been much controversy over the real benefits and costs of both ethanol and the sugarcane used to produce it. Although it is possible to achieve real progress with ethanol on both economic and environmental fronts if the right strategy is chosen, the analysis is highly sensitive to various nuances in the diverse range of feedstocks, fuel blends, productive techniques, distributive strategies, and engine designs. One should keep this in mind when evaluating possible cases. One should also avoid the unproductive tendency to lump all ethanol technologies together.

In this chapter we describe the main characteristics of ethanol as an alternative fuel, addressing some commonly held misconceptions and offering a summary of economic and policy issues as well as an overview of ethanol fuel technology. We then examine the main factors behind the success of the Brazilian sugarcane-based ethanol initiative, addressing not only the structural reasons that favored its adoption and the policy incentives that made it a viable option at inception, but also the transition to a mature market with less governmental intervention. Finally, we explore the implications of sugarcane ethanol viability for selected sugarcane-producing countries in the Americas, including Colombia, Mexico, Cuba, and the United States, in an effort to synthesize possible future trends for fuel energy production and commercialization throughout the world.

6.2 GENERAL CONSIDERATIONS ON ETHANOL

6.2.1 The Technology of Ethanol Fuel

A key factor in the development of the ethanol industry in Brazil was the investment in agricultural research and development by both public and private sectors. Efforts were concentrated on increasing the efficiency of inputs and processes to optimize output per hectare of feedstock and on developing new sugarcane varieties through nitrogen-fixation research to improve cultivation in soils of low fertility. These technological considerations eventually proved crucial to success.

In terms of the production process from sugarcane to ethanol, the typical steps include milling, fermentation, distillation of ethanol, and dehydration. Once harvested, sugarcane is transported in trucks to the milling plant, where it is washed, chopped, and shredded by revolving knives. The feedstock is milled to collect a juice, called *garapa* in Brazil, which contains 10% to 15% sucrose, and *bagasse*, the fiber residue. The main objective of the milling process is to extract the largest possible amount of sucrose from the cane. A secondary but important objective is the production of bagasse with a low moisture content, which is burned for electricity generation, allowing the plant to be energy self-sufficient and to generate electricity for the local power grid. The cane juice is then filtered several times, treated with chemicals, and pasteurized, producing a rich fluid called *vinasse* in the process (Smeets et al. 2006).

Next, evaporation takes place, and the resulting syrup is precipitated by crystallization, producing a mixture of clear crystals surrounded by molasses. A centrifuge is then used to separate the sugar from the molasses. From this point on, the sugar-refining process continues to produce different types of sugar, and the molasses continues on a separate process to produce ethanol. Therefore, the production of sugar ethanol does not necessarily interfere with the production of edible sugar, unlike the corn process; sugar can be turned into ethanol along with the molasses if need be.

In the next step, the molasses is treated to become a sterilized molasses, free of impurities and ready to be fermented. Fermentation transforms the molasses into ethanol through the addition of yeast. This resulting product is hydrated ethanol, used by ethanol-only and flex vehicles in Brazil. Further dehydration is normally accomplished by adding chemicals in order to produce anhydrous ethanol, a substance that is blended with pure gasoline to obtain Brazil's E25 mandatory blend. A typical plant crushes 2 million tons of sugarcane per year, produces 200 million liters of ethanol per year (1 million liters per day over 6 months, April to November), and costs approximately US$150 million. The planted area required to supply the sugarcane is typically 30,000 hectares (Goldemberg 2008).

6.2.2 Four Important Ethanol Fuel Characteristics

Regardless of its source, ethanol has been the subject of ill-placed criticisms. First, ethanol has been criticized because of its lower energy content when compared to gasoline. As shown in Table 6.1, ethanol has a 34% lower energy content by volume than gasoline, meaning that if used in a typical internal combustion engine, ethanol delivers lower fuel efficiency. Some critics argue that this characteristic erodes all the potential practical and environmental benefits of ethanol. However, despite this disadvantage, it has been demonstrated that engines designed for ethanol-only fuel (E100) offer performance equivalent to traditional fuel engines. The higher compression ratios in ethanol as a fuel can allow it to produce more power per unit of displacement if the engine is set up correctly (SwRI 2009). *Insider Racing News* explains:

TABLE 6.1

Energy Content of Different Fuels

Fuel Type	MJ/L (mega joules per liter)	MJ/kg (mega joules per kilogram)	Research Octane Number
Diesel	38.6	45.4	25
Regular gasoline	34.8	44.4	>91
Gasohol (90% gasoline + 10% ethanol)	33.7	47.1	93/94
Liquefied natural gas	25.3	~55	N/A
E85 (85% ethanol, 15% gasoline)	25.2	33.2	105
Ethanol E100 (100% ethanol)	23.5	31.1	129
Methanol	17.9	19.9	123

Source: Davis, S.C. and Diegel, S.W. 2005. *Transportation Energy Data Book: Edition 24.* Center for Transportation Analysis, Oak Ridge National Laboratory. http://www.ornl.gov/~webworks/cppr/y2001/rpt/122271.pdf

Racecars typically utilize engines with reduced piston-to-deck clearance and smaller combustion chambers in the cylinder heads to produce higher compression and higher horsepower. These high compression engines typically require 104 to 110 octane fuels. Pure ethanol produced from plant sources like corn, wheat and sugarcane has an octane rating of 113, which would seem to be a great fit for use in racing. (Madding 2008)

Therefore, it seems clear that ethanol could replace gasoline from a performance standpoint. Naturally, this would require a rethinking of automotive technology and a transition to engines designed specifically for ethanol combustion.

In Brazil, a key technological innovation that greatly aided the transition to ethanol was the development of the flex-fuel engine, which can run on either ethanol, gasoline, or any mixture of the two. Although the first flexible-fuel vehicle was actually Ford's Model T, introduced in 1908, the technology did not evolve due to the domination of fossil fuels in motor vehicles. Given the widespread use of ethanol in Brazil, there was a clear market incentive to provide a cost-effective flexible-fuel engine. The technology was developed for commercial use by the Brazilian subsidiary of Bosch in 1994, but was further improved by 2003 by the Brazilian subsidiary of Italy's Magneti Marelli, located in Hortolândia, São Paulo (de Lima 2009). It uses a computer-based oxygen and airflow sensor feedback system to estimate alcohol content and adjust ignition timing to maximize

TABLE 6.2

Ethanol Fuel Energy Balance

Country	Type	Energy Balance
United States	Corn Ethanol	1.3
Brazil	Sugarcane Ethanol	8.0
Germany	Biodiesel	2.5
United States	Cellulosic Ethanol	2.0–36.0*

Source: Joel K. Bourne, Jr., "Green Dreams," *National Geographic Magazine* 212, 4 (October 2007): 38–59.

* Depending on the production method.

performance. In short, ethanol's lower energy content is a function of the engine and, as described above, the technology exists to overcome this limitation.

A second ill-conceived criticism of ethanol technology is that its production and use result in a net loss of energy. This criticism refers to the energy balance of ethanol, that is, the total amount of energy input into the production process compared to the energy released by burning the resulting ethanol fuel. This is an important consideration when comparing ethanol, which requires an extensive agricultural infrastructure to produce, to gasoline, which results from a directly harvested natural resource. However, it is necessary to keep in mind that different sources of ethanol will result in very different ratios of energy usage. Whereas ethanol from corn leads to small net energy gains, when sugarcane is used to produce ethanol, the process is far more efficient. As shown in Table 6.2, sugarcane provides 8 times as much energy in its resulting ethanol as is required to produce it, compared to just 1.3 times for corn ethanol or 2.5 times for biodiesel. Cellulosic ethanol, which has not been developed to the point of cost effectiveness and commercial viability, can achieve a factor of up to 36. Once developed, however, it would greatly increase the amount of viable feedstock and would allow even fibrous bagasse to be used to produce ethanol, thus providing a streamlined process for sugarcane-ethanol production and opening the possibility of using many other types of raw materials.

A third criticism refers to the technical difficulties in converting some types of raw materials, such as corn, to ethanol. Again, sugarcane-based ethanol has a much higher efficiency than other currently used feedstocks in its conversion to ethanol. As the U.S. Department of Agriculture reports, "Technologically, the process of producing ethanol from sugar

TABLE 6.3

Ethanol Yield per Acre Feedstock

Crop	Ethanol Yield (gal/acre)
Agave (Mexico)	2,000
Miscanthus (U.S.)	1,500
Switch grass (U.S.)	1,000
Sugar beet (France)	714
Sugarcane (Brazil)	662
Cassava (Nigeria and Colombia)	410
Sweet sorghum (India)	374
Corn (U.S.)	354
Wheat (France)	277

Sources: Lester R. Brown, "Beyond the Oil Peak," *Plan B 2.0: Rescuing a Planet under Stress and a Civilization in Trouble* (New York: Norton, 2006). For agave, the source is Sarah Lozanova, "Drink It or Drive It: The Promise of Agave for Ethanol," CleanTechnica.com (accessed August 8, 2008).

is simpler than converting corn into ethanol. Converting corn into ethanol requires additional cooking and the application of enzymes, whereas the conversion of sugar requires only a yeast fermentation process. The energy requirement for converting sugar into ethanol is about half that for corn" (Jacobs 2006). Furthermore, as shown in Table 6.3, sugarcane's ethanol feedstock yield per acre per harvest year is nearly double that of corn. Although some of the other feedstocks lead to higher yields, they are harder to produce in large quantities and are not cost effective.

A fourth (and final) misplaced criticism of ethanol is that its potential to reduce carbon emissions is low relative to other, more sophisticated technologies, such as hydrogen fuel cells, increased electric infrastructure and grid use, and carbon capture and sequestration (CCS), for example. However, the fact that widespread ethanol usage would require a relatively small adjustment to our existing behavior and infrastructure makes it especially attractive. When compared to other climate-change technologies that may mandate drastic changes to automotive engineering, nationwide infrastructures, and large investments in R&D, ethanol's requirements are small. The main necessary changes have to do with engine design, though they are less severe than those required for electric cars or hydrogen fuel cells. In addition, the changeover from gasoline to ethanol at refueling stations would be less costly than setting up the new power sources required for electric cars.

6.3 MAJOR PRODUCERS AND KEY PLAYERS

As the sugarcane ethanol industry grows and develops, it seems most likely that countries which are heavy sugar producers will lead the way. In the Americas, they include Brazil, Mexico, Colombia, the United States, and Cuba (see Table 6.4). Because ethanol facilities must be located in close proximity to sugar plantations to mitigate the high transportation costs, sugar-based ethanol production will be economical only in countries where sugarcane is produced. Currently, only the United States and Brazil have pursued ethanol production on a large scale, and only in Brazil has sugar been the predominant feedstock (see Table 6.5). However, other countries will have large incentives to develop and implement sugarcane-based ethanol technology in the near future.

Colombia, one of Brazil's regional neighbors, has already laid the foundation for its ethanol industry. It is home to one of the world's most efficient sugarcane industries in terms of sucrose yield (tons/acre/year). It shares many of Brazil's environmental characteristics, and the government is eager to help increase its productive capacity of sugar-based ethanol. In

TABLE 6.4

Top Sugarcane Producing Countries (Metric Tons per Year in Millions)

World Rank	Country	Cane Production (million metric tons per year)
1	Brazil	514
2	India	355
3	China	106
4	Thailand	64
5	Pakistan	54
6	Mexico	50
7	Colombia	40
8	Australia	36
9	United States	27
10	Philippines	25
17	Cuba	11

Source: Fischer, G., Teixeira, E. Tothne Hizsnyik, E. and van Velthuizen, H. 2008. Land Use Dynamics and Sugarcane Production. In *Sugarcane Ethanol: Contributions to Climate Change Mitigation and the Environment*, P Zuurbier and van de Vooren, J. (Eds.) 29–62. Wageningen, The Netherlands: Wegeningen Academic. http://www.globalbioenergy.org/uploads/media/0811_Wageningen_-_Sugarcane_ethanol__Contributions_to_climate_change_mitigation_and_the_environment.pdf

TABLE 6.5

Annual Fuel Ethanol Production by Country (in Millions of Gallons)

Country/Region	2008	2007
United States	9,000.00*	6,498.60
Brazil	6,472.20	5,019.20
European Union	733.6	570.3
China	501.9	486
Canada	237.7	211.3
Thailand	89.8	79.2
Colombia	79.3	74.9
India	66.0	52.8
Australia	26.4	26.4
Mexico	14.6	n/a

Source: Renewable Fuels Associates. 2010. Annual World Ethanol Production by Country. Renewable Fuels Association. http://www.ethanolrfa.org/pages/statistics.

* Estimate

addition to Colombia, Mexico has the right climatic conditions to produce sugar and other ethanol feedstocks, such as corn and sorghum, in large quantities. Finally, Cuba is an interesting case because it has the potential to produce huge quantities of sugar, as it did at the turn of the 20th century. As it looks ahead to its post-Castro and potentially post-embargo economic future, ethanol production could become a widespread activity.

At the firm level, most of the key actors in the sugar-based ethanol industry are based in Brazil. Corporate giants such as Cosan Ltd. (NYSE: CZZ) and the Brazilian state-owned oil company Petrobras (NYSE: PBR) lead the way in sugarcane-based ethanol. Petrobras, or Petroleo Brasileiro as the company is officially named, has played a key role in the country's pro-alcohol biofuel promotion program by cooperating with the federal government. In addition, the company plans to spend US$2.4 billion to expand ethanol production to 3.6 billion liters by 2013, of which 1.9 billion will be exported abroad (Hopkins 2009). It has even set aside over half a billion dollars for research and development to ensure that Brazil stays ahead in the biofuel industry. Cosan Ltd. is also progressing through high levels of vertical integration as it acquires new operations in sugar production, ethanol refining, and fuel distribution. After its merger with Nova America, it will account for 10% of Brazil's sugar production and already controls a large part of the fuel distribution network as it purchased ExxonMobil's 1,500 Brazilian service stations (under the Esso brand) in a US$826 million deal completed in 2008. U.S. corn ethanol companies, such as Archer Daniels Midland and VeraSun Energy, did not fare quite as well due to low investor confidence. VeraSun was forced into bankruptcy in 2008.

6.4 THE ECONOMIC AND PUBLIC POLICY ISSUES OF SUGAR-BASED ETHANOL

In theory, sugar-based ethanol promises to impart substantial environmental benefits and fuel independence. In practice, however, the feasibility of implementing this technology comes down to economics. The international market for ethanol is driven by a number of key indicators and prices: the market price of a gallon of ethanol, the production costs of sugar-based ethanol, the production costs of viable feedstocks (sugar, corn, etc.), oil futures and the price of gasoline, and the market price of refined edible sugar. The relationships among these parameters determine the economic feasibility of sugar-based ethanol in a country's domestic market.

Consider first the relationship between the market price of ethanol and related production costs. The value of the market price of a gallon of ethanol will have important implications for the feasibility of producing ethanol from sugar in suboptimal locations where the margins between the market price and the production costs are slim. Places such as the southern United States and many of the Caribbean islands could be considered to be in this category as their sugar production costs are so high that producing ethanol would barely be profitable. Most of the United States would fall in the medium-to-high or high sugarcane production cost classification, with costs around US$.15 per pound of sugarcane (see Table 6.6). Of course, we must also consider the causes of the relatively high production costs and the availability of cost-reducing strategies. If a country's costs are high because of intrinsic factors such as climate or rainfall, they may be harder to reduce than if costs are high because of poor infrastructure, political instability, or out-dated technology.

Another important relative price is that between ethanol and gasoline. A large part of what spurred the high interest in ethanol in the United States were the triple-digit oil prices during the summer of 2008, which drove gas prices above US$4 per gallon. These price hikes increased the demand for, and with it the market price of, ethanol, making it a more viable option for

TABLE 6.6

Costs of Producing Raw Cane Sugar (Cents/Pound), 1994/95–1998/99

Category	Countries	Low	High	Average
Low cost	Central/Southern Brazil, Colombia, El Salvador, and Guatemala	6.72	11.69	7.7
Low-to-medium cost	Florida, Northern/Eastern Brazil, Mexico (Gulf and Pacific Coasts), Costa Rica, Nicaragua, Ecuador, and Bolivia	10.58	17.4	12.34
Medium-to-high cost	Louisiana, Texas, Argentina, Peru, Guyana, Paraguay, Honduras, Panama, and Belize	14.25	21.83	16.54
High cost	Barbados, Dominican Republic, Jamaica, St. Kitts, Trinidad, Hawaii, Uruguay, and Venezuela	17.74	40.21	23.56

Source: LMC International 2000. With permission.

a limited time. Also, because of ethanol's lower energy content, a consumer will choose ethanol only if the price per gallon is around 30% lower than that of gasoline. In the first quarter of 2009, with oil prices hovering above US$50 per barrel, ethanol looked less attractive. However, as China and India demand more and more energy, oil prices are likely to increase in the medium to long term and the potential for ethanol profitability will follow.

One public policy issue directly related to relative pricing is that of import and export tariffs. A current source of tension between the United States and Brazil is the former's tariff of US$0.54 per gallon on all imported ethanol as well as its protectionist policies on refined sugar. From the data in Table 6.6, it is apparent that absent tariff barriers Brazil could easily supply the entire U.S. market for sugar at highly reduced costs, benefiting both countries in the long run. Of course, the United States maintains this policy to protect the development of its less efficient corn ethanol industry and is unlikely to change these policies if the history of agricultural protectionism is any indication of future behavior. The United States should consider, however, that protecting its corn ethanol industry has increased the price of corn and, in the process, harmed its NAFTA partner, Mexico, which depends on the United States for corn supplies now that NAFTA has allowed U.S. producers to push many of Mexico's domestic corn producers out of the market (Dean 2007).

The price of ethanol relative to refined sugar is another important driving force behind the ethanol boom. As has been demonstrated in the Brazilian market, sugar producers will pay close attention to this relationship when deciding whether to refine their raw sugarcane into ethanol or edible sugar. Here we also see the effects of government intervention as, up until the late 1980s, Brazil used direct subsidies to sugar farmers to support ethanol production. However, when these subsidies were removed, sugar farmers switched back to producing refined sugar, leading to a serious ethanol shortage in 1989 (Xavier 2007). Today we see that farmers switch back and forth between refined sugar and ethanol depending on the demand for ethanol and the price of oil.

6.5 THE ROLE MODEL: BRAZILIAN ETHANOL

Brazil's 30-year-old ethanol fuel program is based on the most efficient agricultural technology for sugarcane cultivation in the world. Experts consider its sugarcane ethanol to be "the most successful alternative fuel to date," due to its high market share and productive success (Sperling and Gordon 2009). Using modern equipment and time-tested practices, Brazil has been able to drive production costs below levels seen anywhere else in the world. This has allowed the country's ethanol industry to take full advantage of homegrown sugarcane as a cheap ethanol feedstock. Furthermore, the residual cane-waste (bagasse) is used to process heat and power, which results in a very competitive price for both producers and consumers and contributes to ethanol's high energy balance (output energy/input energy), which varies from 8.3 under average conditions to 10.2 in best-practices production (Rohter 2006). A life-cycle assessment conducted in 2006 by Ekos Brasil showed that, when looking at sugarcane ethanol exported to Switzerland to replace gasoline consumption and thus taking into account the fuel needed to transport the ethanol across the Atlantic, the energy balance was still 5-6:1 (Goldemberg et al. 2008). This means that even when ethanol from sugarcane is exported to distant countries, the final energy balance is very positive.

According to the Renewable Fuels Association, an industry advocacy group, in 2007 Brazil produced 6,472.2 million gallons of ethanol, or 37.3% of the total worldwide ethanol fuel production (RFA 2008). This was second only to the United States. Together, both countries account for 90% of world production of ethanol. However, the fact that Brazilian ethanol production is able to provide for 50% of domestic fuel consumption is unique. Ethanol production in the United States barely makes a contribution to satisfying domestic demand for fuel, which is nearly 25 times that of Brazil. Ethanol's high market share in Brazil has allowed the industry to gain the kind of momentum that will be instrumental in its ability to mobilize resources and increase productive capacity in the future. In fact, an expansion of over 8,000 square kilometers of sugarcane plantations has recently taken place, and Brazil plans to add 77 additional mills by 2013 (Goldemberg et al. 2008).

Brazil's astonishing success, however, could not have occurred without specific pre-existing conditions that allowed the ethanol program to establish itself and progress successfully. It seems as though certain necessary conditions and circumstances existed in Brazil and made it possible for this model to thrive. As mentioned previously, many preconditions have to do with natural endowments, for example, climate and arable land. A well-organized, cost-efficient sugar production system is crucial to the success of sugar-based ethanol. With an average annual rainfall of approximately 160 cm and an average annual temperature exceeding 22°C, Brazil possesses the ideal tropical climate for producing sugarcane (USDA Foreign Agricultural Service 2003). Furthermore, according to the Brazilian Ministry of Agriculture, Brazil has an impressive 320 million hectares of land suitable for cultivation. Currently, only 53 million hectares are under production. Sugarcane accounts for only 5.6 million hectares, or roughly 10% of the total cultivated area. These findings reflect the great potential for expanding sugar production and, thus, sugar-based ethanol. See Figure 6.1 for the geographic distribution of sugarcane plantations in Brazil.

Natural endowments, however, do not tell the entire story. Aside from the favorable production costs for sugarcane, when the ethanol boom began to surge, Brazil had complementary economic and political circumstances that allowed the program to be executed effectively. First, the Brazilian government and its economic policymaking institutions played a crucial role in promoting the ethanol project. As a response to the 1973 oil crisis, the Brazilian government began to support ethanol as a fuel. During the mid-1970s Brazil imported almost 80% of the oil needed to satisfy its domestic demand. At the same time, sugar was commanding low prices in the international market. Taking advantage of these circumstances, in 1975 the government launched a nationwide effort—called the National Alcohol Program (PROALCOOL)—to phase out automobile fuels derived from fossil fuels, such as gasoline, and replace them with sugar-based

FIGURE 6.1
Map of Brazil showing regions where sugarcane is grown as well as environmentally sensitive areas such as the Amazon forest, the Pantanal wetlands, and the Atlantic rainforest. (Goldemberg, José, *Biotechnology for Biofuels*, São Paulo: University of São Paulo, Institute of Electrotechnics and Energy, 2008.)

ethanol. The idea was to use all the excess sugar while simultaneously reducing oil dependency.

This program was financed directly by the government. Initially, it aimed at increasing the number of distilleries in the existing mills, with the federal government offering extremely attractive credit guarantees and low-interest loans for building new refineries. This translated into nearly US$2 billion in loans from 1980 to 1985, representing 29% of the total investment needed (USDA Foreign Agricultural Service 2003). In addition, the government expanded PROALCOOL, using a cross-subsidy scheme whereby gasoline prices were boosted artificially to keep the price of ethanol at a competitive level to promote the production of vehicles designed especially for ethanol use. The Brazilian government also made the blend of ethanol fuel with gasoline mandatory, with the percentage of ethanol fluctuating between 10% and 22% in the period from 1976 to 1992.

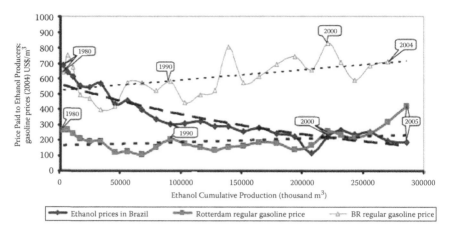

FIGURE 6.2
Evolution of ethanol and gasoline prices to the producer in Brazil. (Goldemberg, José, *Biotechnology for Biofuels*, São Paulo: University of São Paulo, Institute of Electrotechnics and Energy, 2008.)

Figure 6.2 shows the evolution of gasoline and ethanol prices in Brazil over time.

Another important player in Brazil's ethanol success was the state-owned oil company Petrobras. Petrobras not only provided ethanol entrepreneurs with help in transportation, blending and supply, and inventory management, but also provided the government with R&D services and, more importantly, promised to repurchase any excess unsold ethanol fuel. Altogether, the Brazilian government provided three important initial incentives for the ethanol industry: guaranteed purchases by Petrobras, low-interest loans for agro-industrial ethanol firms, and fixed gasoline and ethanol prices where hydrous ethanol sold for 59% of the government-set gasoline price at the pump. These incentives made ethanol production competitive. Without the help of the government and state-owned companies such as Petrobras, the ethanol project would not have been attractive to investors, who played an indispensable role in developing the ethanol industry in Brazil.

Much of this intervention in the form of subsidies and other incentives, however, no longer exists, as market reforms have been implemented to increase competition. During the liberalization of the 1990s, the government gradually rescinded PROALCOOL's incentives and subsidies. Currently, the price difference between gasoline and gasoline mixed with ethanol is defined by the government, and the percentage of the mixture of ethanol with gasoline is set at 25% (Hofstrand 2009). This policy requires close coordination among the various actors involved: the Ministry of Agriculture and the sugarcane planters, the Ministry of Science and Technology and the research centers, the Ministry of Industry and Commerce, the automobile industry, the Ministry of Mines and Energy, Petrobras, the fuel distributors and the gas stations, the Ministries of Finance and Planning, the Ministry of Environment, and automobile owners (World Resources Institute 2008).

Brazil also obtained significant environmental benefits from ethanol development. Several studies have demonstrated that São Paulo benefited from significantly less air pollution thanks to ethanol's cleaner emissions (Green Car Congress 2007). Furthermore, the Sistema Metropolitano de Transporte (Metra—Metropolitan Transport System) has put into operation the first Scania-based E95 ethanol busses. An integral part of this has been the previously mentioned flex-fuel engine technology that takes advantage of the unique characteristics seen in the higher ethanol blends, such as larger oxygen content and higher compression ratios, resulting in lower emissions and improving fuel efficiency.

In sum, the Brazilian ethanol industry's stellar performance is the result of the concerted collaboration of multiple actors, both private and public. Formal policy support began in 1975, and since 1999 prices have fluctuated freely in the market. Most importantly, Brazilian firms and policymakers have amassed a large amount of experience and knowledge in an industry that will continue to grow worldwide. Figure 6.3 depicts the Brazilian energy matrix in 2005 and illustrates the prominent role that sugarcane has achieved.

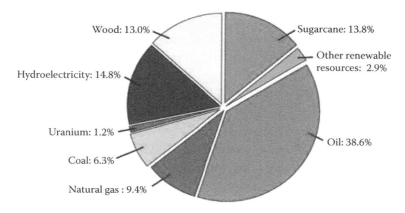

FIGURE 6.3
Brazil's energy matrix in 2005. (Ministério de Minas e Energia do Brasil. With permission.)

6.6 A NEW ETHANOL PRODUCER: COLOMBIA

Over the past few years, Colombia has emerged as the second largest ethanol producer in Latin America, behind Brazil. Colombia began producing sugarcane-based ethanol in October 2005 and produced an average of 277,380 gallons per day in 2007. Its sugarcane is grown in the Cauca Valley, located in the country's central southwest. This area, like the Brazilian sugarcane areas, along the northeast coast and in the State of São Paulo, also benefits from the tropical climate, with an average temperature of 23 to 24°C and an average annual rainfall of 5,000 mm (Toasa 2009).

Currently, Colombia has 518,000 acres of sugarcane plantations, mostly in the Cauca Valley. The valley's total crop area is about 988,000 acres, of which about 53% is used for commercial sugarcane production. About 20% of this land, or 101,000 acres, is used to raise cane for ethanol production. One acre sown with sugarcane in Colombia provides 40 tons of sugarcane and about 740 gallons of ethanol.

The Colombian government, the private sector, and multilateral organizations such as the Inter-American Development Bank, are developing Colombia's potential in order to expand the country's role in the western hemisphere's biofuel industry. Government policies also have helped develop the domestic ethanol industry. In 2001 the government established that gasoline was to contain a 10% ethanol blend by 2006 and a 25% blend within fifteen years. In 2002 Law 788 exempted fuel alcohol from the value-added tax (VAT), and in 2005 and 2006 price supports were established to fix the price of both sugar and alcohol fuel (Rothkopf 2007). Furthermore, Bogota is developing a law requiring that all new cars be equipped to handle fuel with a 20% ethanol blend—the Hacia el E-20 ("Toward the E-20") project—by 2012. It has also created an inter-departmental agency that will oversee more coordinated and sustained development by looking at areas such as prices, taxes, and research (Mackey 2009).

In addition to the government aid, multiple associations are active in the promotion of sugarcane and sugarcane-based ethanol: ASOCAÑA (Association of Sugarcane Growers), PROCAÑA (the Colombian Association of Sugarcane Cultivators and Suppliers), CENICAÑA (the Colombian Sugarcane Research Center), TECNICAÑA (the Colombian Association of Sugarcane Technicians), CIAMSA (the International Commercialization Society of Sugar and Honey), and DICSA (the Industrial Development and Commercial Assembly). Each has specific goals within the industry, such as aiding technology transfer, improving yield, providing logistics for exporting, and helping the industry obtain financing. Combined they make the industry more interconnected and efficient.

Colombia's sugarcane-based ethanol industry, after operating for only three years, is the second-most developed in the western hemisphere. Most of the ethanol plants are energy self-sufficient and even generate surplus power that is sold to the national electric grid. Colombia's sugarcane-based ethanol production is increasing: proposed expansion projects have the potential to more than triple daily production from 277,000 gallons in 2007 to almost 1 million gallons in 2010. Most of this expansion is intended for exports, principally to the United States. This would not have been possible without the help of the government and its supportive industry coalitions, as in Brazil.

6.7 POLITICAL BARRIERS AND MARKET FAILURE: THE UNITED STATES

As the world's leading producer, the importance of the United States in the world ethanol market is undeniable. As of the end of 2008, the U.S. ethanol industry directly supported nearly 500,000 jobs in the 172 ethanol refinery plants in operation, with a combined annual capacity of nearly 10.6 billion gallons (Urbanchuk 2009). Actual U.S. ethanol production in 2008 came to just over 9 billion gallons, compared with around 6.5 billion gallons in Brazil. However, 97% of this ethanol is produced from corn, and there were very few plants in the United States that refined ethanol from sugar. This might be good for the corn producers in the Midwest, but producing ethanol from corn tends to erode the environmental benefits of the resulting ethanol fuel and has serious implications for the price of many corn-based foods. As Joel K. Bourne, Jr., notes, "Corn requires large doses of herbicide and nitrogen fertilizer and can cause more soil erosion than any other crop. And producing corn ethanol consumes just about as much fossil fuel as the ethanol itself replaces" (Bourne 2007). Corn is the most nutrient-intensive crop of all the potential ethanol feedstocks and, therefore, requires a lot of soil inputs to grow. These additional inputs, such as fertilizers and pesticides, inevitably lead to increased pollution and water-table contamination. In the case of corn, the energy balance is not always positive.

Perhaps the most important drawback of corn ethanol is its effect on international food commodity markets through an elevation in the price of corn. This has been, by far, the most controversial issue surrounding the worldwide growth of ethanol technology, as the price volatility of the food we eat is a political issue that has caused many experts and policymakers to look elsewhere for climate-change solutions. In truth, corn prices have a relatively small impact on the overall U.S. consumer price index for food, so to blame ethanol for recent increases in food prices worldwide is slightly unfair and ignores other key factors, such as the increase in the price of oil and other productive inputs (Stebbins 2007). However, it should be noted that in other countries where corn forms a larger part of the diet—including many Latin American countries such as neighboring Mexico—if the corn ethanol industry were to take off in an effort to replace a large portion of U.S. fuel demand, then the impact on food prices could be more significant. In any case, this effect certainly does not help dissipate the doubts about corn-based ethanol's sustainability.

The prospects of mass production of sugarcane-based ethanol in the United States seem remote. Sugarcane can be produced in only four states: Texas, Louisiana, Florida, and Hawaii. A roughly equal amount of sugar beet, which can also be an ethanol feedstock, is produced in the more temperate Pacific Northwest, Great Plains, and Great Lakes regions. However, even when combining both sugarcane and sugar-beet output, U.S. production totals only around 60 million metric tons, which is about one-tenth of Brazil's sugarcane production. Furthermore, sugar production costs are much higher in the United States than in Brazil (see Table 6.6), which drives up the cost of ethanol from sugar and makes corn ethanol a more attractive alternative. In fact, according to a study by the U.S. Department of Agriculture, the costs associated with corn ethanol in the United States are US$1.05 per gallon, whereas those associated with sugar-based ethanol are more than double, at US$2.40 per gallon (Shapouri and Salassi 2006). Brazilian sugarcane ethanol averages US$0.87 per gallon (Bourne 2007).

An equally formidable but perhaps more worrisome force against sugar or any non-corn feedstock is the powerful array of U.S. agricultural lobbies, corn in particular, that stand in the way and often overrun economic and environmental considerations. Robert Bryce, in an article for Yale University's report *Environment 360*, explains how a few realities of American politics and agricultural policy combine to create a perfect storm of corn ethanol subsidies and political gridlock. First, corn is perhaps the most powerful force in the subsidy-rich world of U.S. agriculture. Between 1995 and 2006, approximately US$56.1 billion were spent on federal corn subsidies. According to Bryce, this puts corn subsidies at more than twice the amount given to any other commodity, including American mainstays like wheat and cotton, and over one hundred times more than was paid to tobacco farmers. This is most likely due to the substantial power wielded by the farm state lobbies. Iowa, in particular, is especially notable, given its pivotal presidential caucuses that have forced candidates to make big promises to the corn sector (Bryce 2008). In May 2009, and no doubt due to these powerful political players, the issue of the Environmental Protection Agency raising the mandatory ethanol content of U.S. gasoline blends from 10% to 15% came before Congress. This act would significantly increase the domestic demand for ethanol and be a boon to corn ethanol, but it stalled because of concerns over its effects on older engines in everything from cars to lawnmowers (Jensen 2009). What is also interesting is that even sugar producers seem to shy away from getting behind a sugar-based ethanol movement because of a concern that this may somehow endanger current protective tariffs for sugar.

The most attractive option for the United States to implement ethanol technology effectively seems to be cellulosic ethanol rather than sugarcane- or corn-based ethanol. This technology has not yet advanced to a level where it is ready for commercial production, but in theory it would give the United States extensive freedom in terms of feedstock and capitalize on the comparative advantage that the United States has in high-technology industries. But even though producing sugar-based ethanol in the United States is not optimal, this does not mean that the United States has no role to play in the sugar-based ethanol industry. As the world's leading energy consumer, the United States stands as a potentially irresistible export market for Brazil and any other biofuel-producing nation. In fact, the United States is Brazil's largest market abroad, with imports totaling 2.1 billion liters or 555 million gallons (Licht 2008). If the United States were to adopt ethanol as a fuel source, even if just as a concession to the corn lobbies, this could have important stimulating effects on world ethanol production from all feedstock. Of course, the United States would have to consider reducing its controversial US$.54-per-gallon ethanol import tariff.

This is easier said than done, as agricultural tariffs have become no less than sacred in American politics. However, President Obama seemed to be leaning toward opening up U.S. ethanol production and making clean energy a key ingredient in regional economic recovery. Obama's more outward-thinking policy stance has improved U.S.-Brazilian relations and may provide future opportunities for Brazil to negotiate a reduction in ethanol protection. It is also worth noting, from an economic standpoint, that if the United States were able to perfect the more efficient cellulosic technology-reducing costs to a degree that would make protective tariffs unnecessary, the country would be more willing to open up its ethanol economy. This would benefit still other ethanol-producing nations, as a transition by the United States could encourage a global shift toward ethanol fuel that would dramatically increase demand.

6.8 ETHANOL AND INCOME EQUALITY: MEXICO

Another potential candidate for the ethanol industry and one that would be deeply affected by the U.S. energy policy on this issue is Mexico. Ethanol represents a potential solution to several mounting problems in Mexico, including the exhaustion of domestic oil deposits, rural poverty and marginalization, and environmental degradation (Cantú 2007). According to PEMEX, Mexico's state-owned oil company, the country has proven oil deposits that will last until 2017, with the possibility that surpluses could last a few more decades if efficiency improvements are implemented (Cantú 2007). So, even speaking optimistically, Mexico's domestic oil production will eventually come to an end, and ethanol could help fill the gap in the future to prevent massive oil import bills. Even in the short term, ethanol could fortify domestic energy security by diversifying supply and displacing the 30% of the domestic gasoline consumption imported by PEMEX (Rothkopf 2007).

Ethanol could also help mitigate the country's worsening environmental degradation because the use of biofuels is almost certain to produce on average fewer emissions of greenhouse gases than fossil fuels. A potential first step in a gradual transition to ethanol could be the implementation of a gasoline/ethanol mixture, as in the United States, a measure that could help Mexico save money. As Paul Constance, of the Inter-American Development Bank, explains, in Mexico, "adopting a 10 percent ethanol blend for its domestic gasoline consumption would save nearly US$2 billion per year currently spent to import gasoline and additives" (Constance

2007). Of course, the degree of success of the biofuel model in cleaning up the environment greatly depends on how it is implemented—specifically which technique is used to produce the ethanol fuel. For example, many experts have argued that corn ethanol production in the United States is actually causing a dangerous run-off of aquatic pollutants that is affecting the Mexican economy. These pollutants are increasing the nitrogen levels in the Gulf of Mexico, causing a "dead-zone" algae bloom and negatively affecting aquatic-based industries in Mexico.

Currently, Mexico does not produce a significant quantity of ethanol, but it does produce large quantities of multiple ethanol feedstock—in particular corn, sugar, and sorghum. This occurs despite obstacles such as the poor state of its infrastructure and inefficient land-use practices. These obstacles forced many Mexican corn producers to leave the market after Mexico signed NAFTA and more efficient U.S. producers aided by subsidies were able to undercut them, resulting in the infamous "tortilla crisis." However, thanks to NAFTA Mexico does not face the harmful US$.54-per-gallon U.S. ethanol tariff, as the agreement includes provisions for ethanol and specifies that governments shall not apply restrictions such as tariffs, quotas, and export requirements (U.S. Department of Energy 2006). And any move by the United States to adopt ethanol could promote the use of the technology in Mexico where it is cheaper to produce many of the feedstocks. The most efficient feedstock candidates for Mexico would be sugar and possibly the budding technology of agave, the main ingredient in Mexican tequila. Sugar's more favorable energy balance, combined with Mexico's favorable climate, makes sugar a very attractive option. The southern half of Mexico has a tropical climate and already produces around 50 million tonnes of sugar. In fact, the Mexican sugar industry has been troubled lately by excess supply issues; a sugar-based biofuels industry could help rejuvenate it and create jobs in some of the country's poorest areas. Furthermore, the higher demand for corn in Mexico, due to greater food consumption, nearly all of which it currently imports from the United States, makes corn less viable as an energy source relative to sugar.

Agave, however, may be even more successful than sugarcane and could lead to a real breakthrough in Mexican ethanol. Although more testing and research are needed, preliminary studies show that it enjoys multiple advantages over other options. First, the ethanol yield per acre from agave is much higher than all other feedstocks examined thus far, at nearly 2,000 gallons. Second, because agave is a cactus-like plant, it requires very little by way of water, fertilizers, and pesticides, and grows in harsh desert environments, where most other agriculture would not be possible. Thus, it would avoid two of ethanol's greatest criticisms, as it would not require pesticides and fertilizers that pollute the environment, and it would not compete with other food crops.

Ethanol also has the potential to help the impoverished farmers of Mexico's countryside. However, industry organization will be crucial to ensure that the living standards of the rural poor are improved via an invigoration of the agricultural economy, without inducing price hikes in food staples that will hurt the urban poor. Cantú (2007) stresses that to do this, farmers must capture the rewards of their production and avoid a situation in which they are forced to sell their sugarcane, corn, or sorghum at rock-bottom prices to middlemen who then grab all the upstream profits. He envisions farmer cooperatives setting up their own ethanol mills and dealing directly with distributors. Strong government leadership will also be necessary to avoid predatory practices that will end up hurting rural farmers (see also Leonard 2007).

In 2008 the Mexican Senate took action to stimulate the biofuels industry. The Bioenergetics Promotion and Development Law was passed to empower the use of ethanol from corn and sugarcane. The law gave an incentive to bioenergetics production suppliers, promoted scientific and technology developments in potential crops, promoted the use of integral technology packages, and, furthermore, supported permanent coordination and collaboration with government, political, economic, academic, and social agents (Gómez Macías 2008). It also considered establishing norms to support and advise corn and sugarcane producers in matters of infrastructure, conservation-and-transformation plants, and materials and equipment required for sowing and cultivating both products (Alvarado 2007). This legislation provides a basic framework for the production of biofuels and creates incentives to attract investors.

6.9 A POTENTIAL SUGAR-BASED ETHANOL ENTRANT: CUBA

With a sugar industry that is more than five centuries old and production that stretches across the entire island, Cuba certainly has the basic requirements to take advantage of sugar-based ethanol technology. Just as in Mexico, it seems possible, yet by no means guaranteed, that ethanol could lead to a resurgence of the poverty-stricken rural economy and increase both national income and income equality while helping solve Cuba's energy deficit. In addition, given its low domestic energy demand, Cuba would be able to export a substantial amount of its ethanol and, in the process, improve its trade balance with the rest of the world.

Throughout its history, Cuba has shown that it indeed can produce a lot of sugar. Prior to Castro's communist revolution in 1959, Cuba produced around 50 million metric tons of cane and exported around 5 million metric tons of refined sugar annually, making up almost one-third of global sugar exports (Buzzanell 1992). From the 1960s to the 1980s, Castro's Cuba continued to rely heavily on sugar exports, and its industry benefited greatly from Soviet price supports that allowed Cuban producers to sell their product to the Soviet bloc at highly inflated prices. However, in the early 1990s these price supports vanished as the Soviet Union collapsed, leaving Cuba with an expensive sugar industry that lacked its main customer and was no longer sustainable. Between 1990 and 2000 Cuba's average yield per hectare dropped 38%, and, according to the USDA, per-pound production costs in the mid-1990s were 50% to 70% above world market prices (Peters 2003). In response to the harsh reality that productivity in its sugar industry was too low to support itself, Cuba decided to close factories, convert land use, and divert resources to other pursuits.

However, despite this trend away from sugar, sugar-based ethanol still should be considered a major option for energy production and job creation (Sánchez 2008). A crucial variable to consider here is the cause of Cuba's low productivity. Two likely culprits come to mind. First, Cuba's inefficient land use and resource allocation practices under Castro's communist system collectivized farming into large UBPCs (Basic Units of Cooperative Production) in an attempt to divert profits back to the farmers. We saw a similar pattern in China, with the creation of "people's communes" that provided poor incentives, also leading to low agricultural yields and stifled growth. However, as China demonstrated in the 1980s, modest market reforms and compromises, such as the contract system that served to de-collectivize farming, can reverse this pattern and lead to increases in productivity (Worden et al. 1987). Cuba's strategy of changing what they produce, instead of how they produce it, will not solve its productivity problem.

The second factor contributing to low productivity in Cuban agriculture is the U.S. embargo that has cut off the country from much of the developed world and thus prevented the transfer and absorption of new technologies, not only in terms of hardware but also in terms of software, that is, techniques and ideas. The embargo limits U.S.-Cuban migration and curbs the ability of businesses in the two countries to trade or exchange ideas. This has left Cuba with an infrastructure in disarray and with outdated technologies in its plants and refineries. If Cuba wants to rise again to become a leader in the world markets for sugar and other commodities, it will need to find a way to overcome this technology gap. Certainly, the removal of the embargo in a post-Castro Cuba could greatly aid this pursuit.

The takeaway is that there are no inherent reasons why Cuba's sugar industry is experiencing low productivity and high costs. The true roots of these issues seem to be obsolete productive techniques and outdated government policies, both of which can be reversed or altered. In other words, sugar itself is not the problem. Rather the problem lies in the circumstances surrounding its production. Sugar is still as viable an option as it was for centuries in Cuba, and a move back to sugar and the development of sugar-based ethanol technologies could actually grant the country some advantages over its competitors. Given that its sugar industry is in transition and its ethanol industry would be starting from scratch, Cuba would have the opportunity to implement the newest and most efficient productive technology. This may be possible even if the United States did not open its borders, as Brazil has demonstrated a willingness to share its productive know-how as it tries to make ethanol a leading global commodity. In the event the embargo is lifted and given Cuba's proximity to major U.S. fuel hubs in Louisiana and Texas, Cuban producers would likely enjoy lower transportation costs than those faced by Brazilian producers. Furthermore, as the ethanol market grows and if it were able to replace a substantial segment of fuel demand, this would likely increase sugar prices enough to soften the required reduction in costs by Cuban producers and, in the process, aid in the early phases of sugar-based ethanol development.

Although the economics seem to favor Cuban sugar-based ethanol, tough political barriers remain. In addition to those associated with the U.S. embargo, internal political barriers exist as Fidel Castro has repeatedly and forcefully condemned ethanol, arguing that it will increase both food prices and world hunger. It is likely and understandable that Castro is basing this belief on the early stigma against ethanol, especially corn ethanol, as well as Cuba's experience prior to 1959 in which the sugar industry concentrated wealth in the hands of the few and contributed heavily to rural inequality and unrest. Castro is perhaps right to be wary of ethanol, as Cuba's transformation would likely require heavy outside investment that often leads to unequal distribution of rents and profits. This emphasizes the point that, as in Mexico, the organization of the industry must prevent the exploitation of the farmers. Therefore, Cuba would be wise to pursue a strategy that allows the rural areas to retain profits and that includes government intervention against predatory practices.

6.10 CONCLUSION

The promise of ethanol as a fuel is certainly open to debate. First, ethanol is not a unitary concept, and both the subtle and ostensible differences that arise between different production techniques and applications must be acknowledged. Although many critics tend to lump all ethanol fuels together, it is necessary to understand the differences. This leads us to a second conclusion—that, in the rush to push ethanol technology further, we must not ignore these differences. To ignore the reality that corn ethanol is a problematic and unsustainable technology or that agribusiness must often be controlled to protect farmers would be a critical mistake at a crucial moment.

Unlike corn ethanol, sugarcane-based ethanol is a much more promising technology. It is seven times more efficient to run a car off sugar than off corn. Sugarcane is a significantly less-demanding crop to grow and maintain. Diverting sugarcane to ethanol production will not affect food prices quite as significantly because the sugarcane-to-ethanol process does not consume the edible crystal sugar the way corn does. Also, the bagasse by-product of sugarcane-based ethanol can actually be used to generate electricity for the local power grid. In terms of cost, transitioning to ethanol would require only that consumers spend approximately US$100 to alter their car engines to allow use of an 85% ethanol blend. Other costs, such as those resulting from government subsidies, need to be considered; but, as Brazil has shown, these subsidies can be rolled back after the industry is on its feet. Overall, sugarcane-based technology shows real promise and has the advantage over other carbon-reduction strategies in that it makes good use of existing, rather than new, infrastructure and industries.

Naturally, sugarcane-based ethanol is not feasible everywhere because of climate constraints. Brazil is clearly leading the way; but newcomers such as Colombia, Mexico, and possibly Cuba are likely to develop formidable sugarcane ethanol industries in the not-too-distant future. Industry leaders such as Cosan and Petrobras are looking to expand their operations beyond Brazil. In October 2006 Ecopetrol, Colombia's state-owned energy company, signed an accord with Petrobras for the joint development of biofuel production and distribution systems. The companies agreed to cooperate and conduct studies on production ventures, transportation and infrastructure, and technological support (MEE 2006). In Mexico, in response to government biofuel mandates, the state-owned PEMEX will have to incur extensive capital investments in order to make their factories capable of mixing ethanol with the gasoline they are producing (Pérez 2009).

Certainly, there are other technological options for reducing our impact on the environment, staving off climate change, and securing fuels. Several more advanced technologies—such as hydrogen fuel cells, solar power, carbon capture and sequestration, solid-state lighting, and superconductivity—offer even greater potential gains in terms of carbon reduction and sustainability. However, ethanol possesses a few major advantages over these alternatives, including low transition requirements, demonstrated commercial success, viable business models, and political momentum. Given the need for swift action against global warming, ethanol may be the best option on the table at the moment. Furthermore, the sugarcane-based ethanol variant also contains the optimal balance of feasibility and the best carbon-reducing power of the various ethanol alternatives. If implemented, ethanol will likely buy us much-needed time to research and develop the untested, more advanced, and higher-return technologies.

7

The Challenge of Sustainable Tourism

Jessica Webster

CONTENTS

7.1 Introduction...135
7.2 Ecuador and the Galapagos Islands 138
    7.2.1 Characteristics of the Galapagos 140
    7.2.2 Human Involvement with the Galapagos...........................142
7.3 Tourism and Tourists in the Galapagos 150
    7.3.1 Overview 150
    7.3.2 Stakeholders...152
    7.3.3 Current Concerns 153
7.4 Main Issues as Expressed by Stakeholders in the Galapagos..........155
    7.4.1 Awareness of the Current Situation in the Galapagos........156
    7.4.2 Tourists' Central Concern When Traveling to the
           Galapagos Is Cost...156
    7.4.3 Operators' Choice to "Be Sustainable" Still Boils Down
           to Profitability and Ethics 158
    7.4.4 Ecuadorian (Domestic) Tourist Vision Misalignment........159
    7.4.5 Tensions between Tour Operators and Retailers in the
           Galapagos Islands 160
    7.4.6 A Negative Cycle: Tourism Driving Migration, Driving
           Expansion, Driving Destruction 161
    7.4.7 Better Education and Stronger Law Enforcement................162
7.5 Conclusions...162

7.1 INTRODUCTION

Tourism is one of the largest sectors of the economy, representing between 5% and 10% of employment and gross domestic product in most countries. In smaller, tourism-oriented island-nations, it can represent upward of 60%. It should be no surprise that the World Tourism Organization has estimated that tourism directly or indirectly accounts for about 5% of global carbon emissions. Moreover, its impact on fragile ecosystems, like those of coastal areas, can be significant and detrimental. As tourism continues to grow, it is important to understand its impact on destination environments, communities, and economies—and to develop a form of leisure that is sustainable and mutually beneficial for all parties and players.

International tourism has been increasing dramatically and is projected to more than double from 694 million international tourists in 2005 to 1.6 billion in 2020, of which 1.2 billion are expected to travel within their own regions, while the rest are expected to travel between regions—resulting in annual growth rates for regional travel of approximately 3.8% and for interregional travel of 5.4%. Europe currently accounts for the largest share of tourist arrivals with 53.1%, followed by Asia and the Pacific at 20%, the Americas at 15.9%, the Middle East at 6.0%, and trailed by Africa at 5.1%. The average annual growth rates of regional destinations are predicted to be greatest in the Middle East (6.7%), East Asia/Pacific (6.5%), and South Asia (6.2%), with the lowest in Europe (3.1%) and the Americas (3.8%) (UNWTO 2009).

Tourism plays an important role in the global economy. The industry generated US$1.1 trillion in revenue during 2008, or approximately US$3 billion per day. It is also the world's fourth largest export category, accounting for as much as 30% of global exports of commercial services and 6% of overall exports. Furthermore, many developing countries have come to rely on tourism as one of their main sources of income (UNWTO 2009).

The World Tourism Organization breaks tourism down into four categories according to the reasons for travel. The first is "leisure, recreation, and holiday," which makes up the largest portion and accounts for 51% of the total; "VFR [visits to friends and relatives], health, religion, other" represents 27%; "business and professional," 15%; and lastly, "not specified," 7%. Each of these categories contributes to environmental degradation because of the travel and accommodation required. The majority of tourists travel by air (52%) and land (42%)—which can be further subdivided into road (39%) and rail (3%). Very few travel by water (6%) or other means (4%). Air travel continues to grow at a faster rate than land transport, thereby gradually increasing its share over time. Air travel is particularly troublesome in terms of its environmental effects. Aviation, in general, contributes to approximately 3% of global carbon emissions (Rice-Oxley

2007), which will, of course, increase as the rates of both tourism and air travel increase (UNWTO 2009).

As tourism continues to grow, carbon emissions from flying are not the only side effect of concern. Tourism has implications for communities, the environment, the culture, the flora, the fauna, the economy, and so on. A tourist *always* has an impact. The questions then become, with so much potential for damage or change: (1) how can tourism's negative effects be mitigated or eliminated, and (2) how can the visited communities and their environments best be supported and/or protected? The term "sustainable tourism" has been coined as a strategy to address this challenge, with a commonly cited definition from the World Tourism Organization:

> Sustainable tourism development meets the needs of the present tourists and host regions while protecting and enhancing the opportunity for the future. It is envisaged as leading to management of all resources in such a way that economic, social and aesthetic needs can be fulfilled, while maintaining cultural integrity, essential ecological processes, biological diversity and life support systems. (Manning et al. 1997)

Sustainable tourism is not a "type" of tourism as, for example, "adventure" and "luxury" are. Rather, it is a manner of traveling or a way of designing tourism so that present conditions may be preserved or improved for the future. Furthermore, an important distinction exists between sustainable tourism and the currently hot buzzword, "eco-tourism." Eco-tourism relates to protecting the *environment* from tourists, without mention of the community or cultural preservation; it is "tourism that involves traveling to relatively undisturbed or uncontaminated areas with the specific objective of studying, admiring, and enjoying the scenery and its wild plants and animals, as well as any existing cultural manifestations (both past and present) found in these areas" (Ceballos-Lascurain 1991, 25). Thus, sustainable tourism is a much more encompassing concept.

Tourism has a unique capacity to either help build or help destroy resources and opportunities for present and future societies and, as such, it is necessary to understand the practical incentives and implications on a case-by-case basis, that is, at the local level. The opportunities for positive impact are immense because of the scope and scale of this economic activity, but it is an area in which generalizations can be dangerous. As pointed out by Budeanu (2005), to verify how tourism can contribute to sustainable development, it is necessary to understand the points of view of the various actors in the industry in a given region. Tepelus (2005) underlines the importance of tour operators in the tourism supply chain, stresses the need to disseminate good practices, and provides specific examples of such sustainable tourism initiatives. Choi and Sirakaya (2006) used the Delphi method to develop a broad set of sustainable tourism indicators that can be used as a starting point for understanding issues at the regional and local levels. The literature on sustainable tourism is divided into two main categories of inquiries. One group is mostly descriptive and focuses on successes and failures of observed initiatives, whereas the other is mostly normative and prescriptive, focusing on the "ideal."

This chapter seeks to contribute to the debate by examining the state of sustainable tourism in one of the world's most iconic ecosystems, the Galapagos Islands. This archipelago located 600 miles off the coast of Ecuador is an ideal candidate for examining tourism through the critical lens of sustainability for three main reasons. First, the archipelago is isolated and, as such, it provides a unique "laboratory" to understand the impact of tourism on the environment. Second, tourism is the main activity on the islands and, therefore, the impact of tourism can be observed without interference of other economic activities—furthermore, the social consequences of tourism initiatives can be identified explicitly. And third, tourism in the Galapagos has grown exponentially over the past 25 years, leading to an increased realization on the part of virtually every stakeholder of the urgent need to preserve its unique environment for the future. The aim of this chapter is not to offer a solution to each of the challenges outlined, but rather to highlight thoughts and perspectives of different local stakeholders with the objective of finding the balance among environmental, social, and economic sustainability initiatives.

7.2 ECUADOR AND THE GALAPAGOS ISLANDS

Ecuador is a popular and growing tourism destination, with a recent annual foreign tourist arrival growth rate of approximately 7%. In 2008 slightly more than one million international tourists visited the country. While tourism arrivals increased by 20% between 2006 and 2008, revenue witnessed an increase of 55% (UNWTO 2009). The country markets itself as being *el país de cuatro mundos*, the country of four worlds: the coast, the Andes mountains or highlands, the Amazon, and the Galapagos Islands.

The coast is where one would go for beaches and relaxation. In the Andes, one can go trekking and climbing on some of the highest mountain peaks in South America, several over 5,000 meters and covered in year-round snow. In the Amazon, one can experience the jungle and the indigenous communities of the rain forest. And the Galapagos Islands offer unparalleled exposure to an animal-viewing paradise in a volcanic terrain like no other on Earth. These are truly four different worlds a tourist can explore when visiting Ecuador.

Ecuador's Ministry of Tourism was established in 1992 and since then, with mixed success, has tried to pass legislation, organize programs, and issue certifications to help attract tourists and to protect the environment and the communities. Their current focus is on the PLANDETUR 2020 [Tourism Plan for 2020], which has the following three main stated objectives:

1. To initiate a process that coordinates the public, private, and community strengths for the development of sustainable tourism, based in their territories and under the principles of alleviating poverty and supporting equality, sustainability, competitiveness, and decentralized management.
2. To create conditions that allow for sustainable tourism to be a basis for speeding up the Ecuadorian economy, looking to improve the quality of life of the population and the satisfaction of the actual tourism demand—taking advantage of Ecuador's comparative advantage and unique characteristics.
3. To insert sustainable tourism in the policies of the state and in national planning to promote complete development and rationalization of public and private investment (Ecuadorian Ministry of Tourism 2007).

Sustainability—not only of nature, but also of the indigenous communities, the economy, and the tourism industry itself—is being targeted explicitly. The Ministry of Tourism is also making a push toward improving education for those in the tourism industry, such as the National Program for Tourism Training, which was launched in October 2009. Furthermore, the Ministry is actively involved in bettering the lives of micro-, small-, and medium-sized entrepreneurs, such as those involved with maritime transportation (e.g., My Touristic Canoe), as well as with integrating taxi service into the tourism industry through accreditations and certifications in hopes of improving the quality of service and information to tourists (e.g., Info Taxi).

7.2.1 Characteristics of the Galapagos

The Galapagos Islands—an archipelago consisting of 13 large islands, 6 smaller ones, and over 40 islets—are located approximately 600 miles (1,000 kilometers) off the coast of Ecuador in the Pacific Ocean (see Figures 7.1 and 7.2). They are volcanic in nature, having begun their emergence from the sea 3 to 5 million years ago and continuing to erupt, grow, and shift to this day. The eastern islands, such as Española, are the oldest, while the youngest, such as Fernandina, have rocks indicating they are less than 200,000 years old. The islands' land mass is approximately 3,042 square miles (7,880 square kilometers), with a maximum height of 5,600 feet (1,707 meters). Only five of the islands are inhabited by humans.

The Galapagos are unique because of their volcanic formations, but more so because of their isolation from the rest of the world. This isolation has resulted in some of the highest levels of endemism of any location

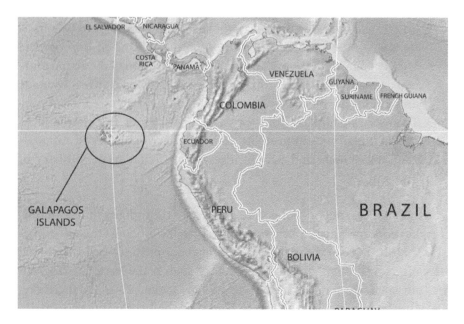

FIGURE 7.1
The Galapagos Islands in relationship to North and South America. (Central Intelligence Agency, http://www.cia.gov, accessed December 13, 2010.)

FIGURE 7.2
Satellite view of the Galapagos Islands. (Based on the public domain NASA satellite photo image: Galapagos-satellite-2002.jpg.)

on the planet, meaning that a large number of species are native to the Galapagos and cannot be found anywhere else on Earth (Darwin 1859; WWF 2009). Over 9,000 species live on the islands or in the surrounding waters, including over 400 species of fish. Three-fourths of all reptiles and land birds are endemic; and of the 500 to 600 indigenous plant species, such as the black mangrove, beach morning glory, and the lava cactus (see Figure 7.3), approximately one-third are endemic (INGALA 2009; WWF 2009). Furthermore, the Galapagos are home to the only penguins found north of the equator (see Figure 7.4) and the only lizards (marine iguanas) that swim in the ocean (see Figure 7.5). There is also a distinct lack of natural predators on the islands, and the animals are both unaccustomed to and unafraid of humans. The Galapagos are indeed like no other place on Earth.

Furthermore, the Galapagos' isolation has also provided the opportunity to see the effects of evolution and natural selection on animals and plants. Because the islands are very close together, one might imagine that species would frequently travel, migrate, or even visit different islands. However,

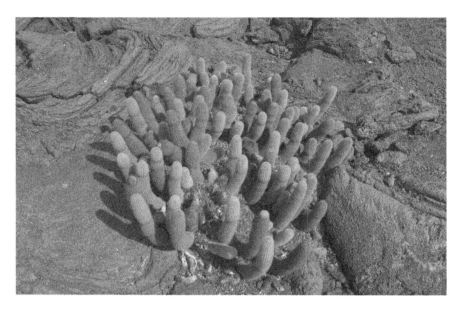

FIGURE 7.3
Lava cactus on the Galapagos. (Photo Jessica Webster. With permission.)

this has not been the case. Many of the various islands' populations have actually remained quite isolated from each other, which means not only that some species are endemic to the Galapagos, but also that several species are endemic to particular *islands.* As Charles Darwin explained in his 1859 *Origin of Species,* "In the Galapagos Archipelago, many even of the birds, though so well adapted for flying from island to island, are distinct on each [island]; thus there are three closely-allied species of mocking-thrush, each confined to its own island." Different types of island landscapes, flora, and fauna, including sandy beaches, rock formations, open lava fields, tide pools, forests, mangroves, salt water lagoons, cracked lava formations, and lava tubes, are found in the Galapagos (see Figures 7.6 through 7.15 for some examples).

7.2.2 Human Involvement with the Galapagos

The Galapagos were first discovered, uninhabited by humans, in 1535 by the Bishop of Panama and were mainly visited only by pirates until the end of the 18th century, when whalers and fur traders arrived (INGALA 2009). Ecuador annexed the Galapagos in 1832, but Darwin's research during 1835 bears more responsibility for bringing the islands to the attention

FIGURE 7.4
Galapagos penguin, the only penguin found north of the Equator. (Photo Jessica Webster. With permission.)

of the world. Not much changed in the islands over the next one hundred years—little migration and few visitors—although those who did make their way were likely "conscripted vagrants, political dissidents, and prisoners condemned to one or another of the notoriously inhumane penal colonies that existed at various times on Floriana, San Cristóbal, and

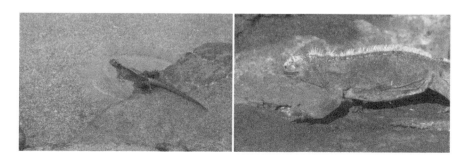

FIGURE 7.5
Ocean-swimming marine iguanas. (Photo Jessica Webster. With permission.)

FIGURE 7.6
San Cristobal: sandy beaches and seals. (Photo Jessica Webster. With permission.)

Isabela" (Epler 2007, 2). The islands were also used as military outposts at various times.

The Galapagos remained isolated until the advent of jet aviation. The late 1950s proved transformational for the islands. People began to immigrate to the four colonized islands, the Charles Darwin Foundation was established for research and conservation of the islands, and the Galapagos National Park was created (Graves 2006, 3). Further protection was eventually established—UNESCO named the islands as a World Heritage Site in 1978, which was extended to the Marine Reserve in 2001, and also recognized the Galapagos National

FIGURE 7.7
Rock formations: Kicker Rock. (Photo Jessica Webster. With permission.)

FIGURE 7.8
Genovesa: red-footed boobies, male frigate birds (red pouches). (Photo Jessica Webster. With permission.)

Park as a Biosphere Reserve in 1984. During the 1960s immigration continued to rise, and by the end of the decade and the early 1970s organized tourism did as well (Epler 2007, 3). During the 1970s the number of tourist vessels grew from 4 or 5 to 40, although the number of tourists remained fairly stable from 1980 to 1985, averaging 17,500 per year. This number, unfortunately, was already higher than the officially recommended guidelines of 12,000 visitors per year (Epler 2007). After 1985, tourism skyrocketed, more than tripling to 41,000 in 1990, almost doubling again to 72,000 in 2000, then rising by an additional 70% to reach 122,000 tourists in 2005 (Epler 2007) and rising yet again to more than 170,000 in 2008 (Galapagos National Park 2009). During this time, immigration continued as well. The population of the islands stood at over 28,000 in 2008 (WWF 2009). Figure 7.16 reflects the

FIGURE 7.9
Fernandina: salt-water lagoons and tide pools. (Photo Jessica Webster. With permission.)

FIGURE 7.10
Fernandina: Sally Lightfoot crabs, flightless cormorant (Look, no wings!). (Photo Jessica Webster. With permission.)

FIGURE 7.11
Santiago: cracked lava formations and the Chinaman's Hat. (Photo Jessica Webster. With permission.)

FIGURE 7.12
Bartolome: moonscape of a once-active volcano. (Photo Jessica Webster. With permission.)

FIGURE 7.13
North Seymore: endemic land iguanas and blue-footed boobies. (Photo Jessica Webster. With permission.)

FIGURE 7.14
Santa Cruz: giant tortoises. (Photo Jessica Webster. With permission.)

FIGURE 7.15
Española: waved albatross and the "famous blow hole." (Photo Jessica Webster. With permission.)

exponential increase in tourism and immigration, and Table 7.1 gives the numbers of entries broken down by nationals and foreigners. Both immigration and tourism have increased steeply since 1990. Whereas during the 1990s immigration grew faster than tourism arrivals, during the 2000s the latter grew more.

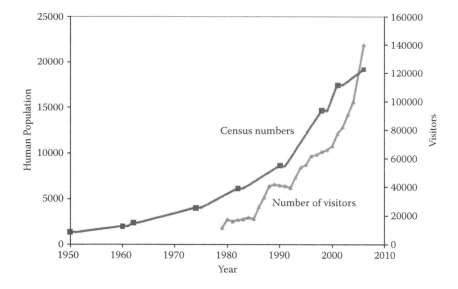

FIGURE 7.16
Growth of the population in Galapagos and numbers of visitors to the Galapagos. (G. Watkins and F. Cruz, *Galapagos at Risk: A Socioeconomic Analysis of the Situation in the Archipelago,* Puerto Ayora, Galapagos, Ecuador: Charles Darwin Foundation, 2007. With permission.)

TABLE 7.1

Entries to the Galapagos by Ecuadorean Nationals and Foreigners since 1979

	Nationals	Foreigners	Total	Nationals (% Total)	Foreigners (% Total)
1979	2,226	9,539	11,765	18.92%	81.08%
80	3,980	13,465	17,445	22.81%	77.19%
81	4,036	12,229	16,265	24.81%	75.19%
82	6,067	11,056	17,123	35.43%	64.57%
83	7,254	10,402	17,656	41.09%	58.91%
84	7,627	11,231	18,858	40.44%	59.56%
85	6,279	11,561	17,840	35.20%	64.80%
86	12,126	13,897	26,023	46.60%	53.40%
87	17,769	14,826	32,595	54.51%	45.49%
88	17,192	23,553	40,745	42.19%	57.81%
89	15,133	26,766	41,899	36.12%	63.88%
90	15,549	25,643	41,192	37.75%	62.25%
91	14,815	25,931	40,746	36.36%	63.64%
92	12,855	26,655	39,510	32.54%	67.46%
93	10,136	36,682	46,818	21.65%	78.35%
94	13,357	40,468	53,825	24.82%	75.18%
95	15,483	40,303	55,786	27.75%	72.25%
96	16,113	45,782	61,895	26.03%	73.97%
97	13,979	48,830	62,809	22.26%	77.74%
98	14,440	50,351	64,791	22.29%	77.71%
99	12,602	53,469	66,071	19.07%	80.93%
'00	14,561	54,295	68,856	21.15%	78.85%
'01	19,711	57,859	77,570	25.41%	74.59%
'02	22,939	59,287	82,226	27.90%	72.10%
'03	28,346	62,999	91,345	31.03%	68.97%
'04	33,977	74,957	108,934	31.19%	68.81%
'05	35,584	86,110	121,694	29.24%	70.76%
'06	47,833	97,396	145,229	32.94%	67.06%
'07	51,411	110,448	161,859	31.76%	68.24%
'08	53,468	119,952	173,420	30.83%	69.17%

Source: Galapagos National Park, http://galapagospark.org/Informe_ingreso_turistsas_2008.pdf (accessed November 5, 2009).

7.3 TOURISM AND TOURISTS IN THE GALAPAGOS

7.3.1 Overview

The environmental impact of tourism on the Galapagos depends on several factors, the most important of which is the increase in number of visitors, as described in the previous section. There are several different categories of visitors to the islands. Tourists arrive at the archipelago and then either visit one or more islands on all-inclusive live-aboard vessels, or stay in hotels visiting islands on local boats. Category 1 tourists (31%) are the wealthiest, oldest segment and tend to stay on larger, more expensive boats; Category 2 tourists (43%) have a slightly lower average income and age, and lean toward smaller 16- to 20-person boats; and Category 3 tourists (7%) are mainly backpackers and students on vacation for more than 7 days (Epler 2007, 9). More detail on these categories is presented in Table 7.2.

Although visiting the Galapagos by boat or by hotel may not sound mutually exclusive, there actually is an undeniable polarization. A striking difference exists between how Ecuadorian tourists experience the islands and how foreigners do: 81% of Ecuadorian tourists stay *only in hotels* on the islands, while 93% of the Category 1 tourists and 59% of Category 2 tourists stay *only on boats*. Only 7% of Category 1, 17% of Category 2, 34% of Category 3, and 5% of Ecuadorian tourists stay both on boats and in hotels (see Table 7.2). Also, it should be noted that although mainland Ecuadorians account for approximately 30% of all those entering the Galapagos, they account for only 18% of entrants who come for *tourism* reasons. While the number of passenger vessels has been relatively steady since 1997 (80–84), the passenger capacity (total and per vessel) has continued to increase slowly (see Table 7.3)—reaching in 2006 a total capacity of 1,805 and passenger capacity per vessel of 22.6 (Epler 2007, 18; Galapagos National Park 2009).

Tourists have various ways of planning their trips and arriving in the Galapagos. Many book their tickets while still in their home countries, whereas others, who are less risk-averse or have fewer demands on their time, may book after arriving in Ecuador. Both Quito, the capital of Ecuador, and Guayaquil, an industrial coastal city, have tourist areas littered with travel companies looking to provide "last-minute deals" to travelers—most of which are geared to a slightly lower end, for example,

TABLE 7.2

Characteristics of Galapagos Tourists

Profile Characteristic	Foreigner Category 1	Foreigner Category 2	Foreigner Category 3	Ecuadorian Category 4
Average Age	50	44	28	38
Income Distribution				
50,000/Yr or less	13%	26%	56%	
$50,000–$75,000/Yr	21%	32%	38%	
$100,000/Yr plus	66%	42%	6%	
Number of travel companions paid for	2.8	2.4	1.5	3.5
Origin				
North America	60%	39%	42%	NA
Europe	29%	48%	34%	NA
South America	2%	5%	2%	NA
Other	9%	8%	22%	NA
Booked a package tour	77%	68%	50%	33%
Percent Staying In				
A vessel only	93%	59%	26%	14%
A vessel and hotel	7%	17%	34%	5%
A hotel only	NA	24%	32%	81%
Private residence/with families				19%
Nights in Galapagos Spent on				
A vessel	5.7	4.3	4.1	0.9
A hotel	0.5	2.0	3.6	2.8
Private residence/with families				1.5
Total days in Galapagos	6.2	6.1	7.7	5.2

Source: Bruce Epler, *Tourism, the Economy, Population Growth, and Conservation in Galapagos* (Puerto Ayora, Santa Cruz: Charles Darwin Foundation, 2007, p. 10. With permission.)

backpackers or extended-time travelers. It is cheaper to buy tours once in Ecuador because there is often one less middleman. If tours are purchased from abroad, many times there is a front-end travel agent in the home country, whereas in Quito or Guayaquil, one deals directly with the actual service provider.

Although one can arrive in the Galapagos by either air or sea, the vast majority of travelers arrive by air. Three commercial airlines (AeroGal, Tame, and LAN Ecuador) fly from Guayaquil and Quito to the Galapagos'

TABLE 7.3

Number of Tourist Vessels in the Galapagos and Their Capacity

Year	1981	1991	1996	1997	2000	2006
Number of vessels	40	67	90	84	80	80
Total passenger capacity	597	1048	1484	1545	1733	1805
Passenger capacity/vessel	14.9	15.6	16.5	18.4	21.7	22.6

Source: Bruce Epler, *Tourism, the Economy, Population Growth, and Conservation in Galapagos* (Puerto Ayora, Santa Cruz: Charles Darwin Foundation, 2007, p. 18. With permission.)

two commercial airports, one on San Cristóbal and the other on Baltra, several times daily. Most tour companies will include the cost of the flight (round trip from Quito or Guayaquil to the islands) in the tour package. For foreigners this cost is between US$334 (low season) and US$390 (high season). For continental Ecuadorians, the flights cost less and are often bundled with a hotel package.

7.3.2 Stakeholders

For a small, remote set of islands, the Galapagos have an extensive list of stakeholders:

- **Residents of the islands:** The majority of those living on the islands have moved there because of tourism. Wages are higher in the Galapagos than on the mainland—around 70% higher on average according to Patel (2009), making it an attractive option.
- **Tourists:** As discussed above, tourists have an obvious stake in the Galapagos. Foreign tourists seek the idyllic paradise described by Darwin and are willing to pay a lot to experience it. Whereas Ecuadorian tourists tend to stay in hotels and enjoy the sun and beaches, foreign tourists usually prefer to stay on boats.
- **Tourist companies, travel agencies:** As trite as it sounds, tourism in the Galapagos is what sustains these companies and, as a result, they should be invested in maintaining and expanding the islands' potential. Their ability to earn revenue *in the long run* is contingent upon the preservation of islands and wildlife. These companies are also subject to state and National Park regulations and quotas.
- **Conservation groups:** The Charles Darwin Foundation, created in 1959, the World Wildlife Fund, and the Galapagos Conservancy

(based in the United States) are arguably the largest and most important conservationist groups focused on the Galapagos. They work actively to protect the islands and animals via research and publications, develop policies and strategies for sustainability, execute joint projects with multinational philanthropic corporations and research institutions, and promote education/awareness (among tourists and locals as well as globally). The Darwin Foundation is located on the island of Santa Cruz and also contains a breeding ground.

- **Donors/philanthropic international corporations:** Green is "hot" in the early 21st century and some companies, such as Toyota and Motorola, have an interest in the Galapagos from a philanthropic point of view as well as, one would assume, from a marketing angle. They partner with the conservation groups and travel companies (often offering technology and know-how) to support sustainable tourism and lifestyle practices in the Galapagos and to protect the National Park.

- **Government agencies:** INGALA (the Galapagos National Institute), the Galapagos National Park, and the local government are the major governmental players in determining policy and regulations within the Galapagos (such as migration and visiting restrictions) and for protecting the sustainability of the islands, of residents, and of tourism. INGALA was created in 1980 and was further strengthened under the "Special Law for Galapagos" in 1998. It is responsible for "coordinating regional planning, government funding (national, bilateral and multi-lateral assistance), and technical assistance in Galapagos" (INGALA 2009). Founded in 1959, the National Park is responsible for the conservation and integrity of the ecology and the biodiversity of the islands—ensuring the sustainability of the goods and services generated by the islands for the community (Galapagos National Park 2009). The Ecuadorian Ministry of Tourism also plays an overarching role in tourism for the country as a whole and continues to develop all-encompassing legislation and strategies such as the Tourism Plan for 2020. Certifications such as SmartVoyager have also been launched through the government as protective measures.

7.3.3 Current Concerns

The Galapagos face many current interrelated challenges, most of which boil down to how people (tourists, residents, and tour operators alike) and introduced (i.e., non-indigenous) species of plants and animals affect the natural habitat and the long-term sustainability of the archipelago. Four main categories of concerns seem most relevant:

- **The effects of introduced plants and animals on the ecosystem:** Many non-indigenous species of plants and animals have arrived or have been brought to the islands, sometimes unwittingly by tourists, over the years, compromising a significant number of native plant and animal species. A 2007 report suggests that 60% of the endemic plant species are threatened (according to the IUCN Red List criteria), mainly because of the impact of development and of introduced herbivores (Atkinson 2008). Two methods for dealing with introduced herbivore species in order to protect vegetation are (1) island-wide eradication, and (2) fencing in individual populations. Pigs, donkeys, and goats are examples of introduced species that certain islands, such as Santiago, have successfully eradicated. Introduced plant species, such as the blackberry, also have degrading effects on the environment (Atkinson et al. 2008). Given the harmful effects of introduced plants and animals, two of the Galapagos' most important challenges are (1) to reduce/limit—through quarantine or waste-removal strategies, etc.—the arrival of these foreign flora and fauna, and (2) to continue eradicating and/or containing the present invading species.
- **Effects of tourism on the animals and the environment:** There is no denying that tourism impacts the environment:
  - **Tour operators:** While rules and regulations have been designed to protect the environment, especially with regard to tour operators and waste and water management, there appears to be a lack of enforcement and compliance. As the 2007–2008 Galapagos Report explains, "The tourist fleet of Galapagos, taken in aggregate, is deficient in its environmental compliance and is even less active in ensuring best environmental practices in areas that are unregulated" (Guime and Muñoz 2008). This leads to increased pollution, fuel use, the introduction of contaminating species, and more—all of which harm the environment and its sustainability.
  - **Tourists:** When visitors walk on the islands and dive/snorkel in the waters, they can have a negative impact in several ways. For example, visitors may destroy the ecology by stepping on or touching nature, especially as the tourist numbers increase and

the same pathways are trod upon. Tourists can also change the behaviors and reactions of animals with which they come in contact (Cubero-Pardo and Bastidas 2008). As tourism increases, the impact will be increasingly negative.

- **The challenge of migration/population control:** The fact that more and more people continue to migrate from the mainland poses a sustainability problem. They consume resources (e.g., food, water, electricity, fuel), need jobs, require homes, etc. They also continue physical expansions of towns and tourism infrastructure onto more of the natural habitat—again at a sacrifice to the environment. For example, as of October 2009, 200 new cinderblock homes were being built on the outskirts of Puerto Villamil, Isabel's main town, and a project called El Mirador was clearing space for another 1,000 (Romero 2009). To partially combat population growth rates, there is a new trend in deporting mainland Ecuadorians from the islands—normally the poor who have come and stayed illegally. It is estimated that approximately one-fifth of the Galapagos population is illegal (Patel 2008). Between January and October 2009, approximately 1,000 were returned to the mainland (Romero 2009).

- **Fishing and fisheries:** Although there has been a substantial amount of research and regulation around fishing and fisheries worldwide, because fishing is not a major tourism driver in the Galapagos, it is not addressed in this chapter. For more information on this topic, please see CDF, GNP, & INGALA 2008; Graves 2006; Epler 2007.

7.4 MAIN ISSUES AS EXPRESSED BY STAKEHOLDERS IN THE GALAPAGOS

This section explores the Galapagos through the eyes of its stakeholders. Seven different travel agencies in Quito, fourteen tourists on a small-sized vessel in the Galapagos, a development officer at the Charles Darwin Foundation, five shopkeepers in the Galapagos, an Ecuadorian environmental lawyer, and an employee at the Ministry of Tourism in Quito contributed their thoughts on the current state of the Galapagos, including challenges and recommendations for improvement. Although much of what these individuals discussed is reflected in the literature, these on-the-ground interviews shed new light on issues unrepresented or underrepresented in current Galapagos research.

7.4.1 Awareness of the Current Situation in the Galapagos

Travel agents and the development officer at the Charles Darwin Foundation, Roslyn Cameron, explained that most tourists assume that enough is being done to protect the islands and that current challenges are being met. Tourists believe that the Galapagos Islands "are protected" already, missing the fact that protection is a continuous process, involving money, time, research, and continuous efforts on the part of all stakeholders. Interestingly, this mentality differs from tourists' associations with the rainforest/jungle, where they are more conscious of the impact of tourism and assume they play a more active role in conserving the rainforest—they know that being "eco-conscious" is important. As Kimrey Batts, of the Happy Gringo travel agency, commented, "With the rainforest, it's something our generation has grown up with thinking that it's threatened. There's a 'sense of threatened' related to it. The rainforest is also associated with communities and the concept of protective measures. Unfortunately with the Galapagos, there's no sense of danger or threat for a lot of people. They just assume it's being taken care of." Tourists must be better informed about the current situation and the islands' needs.

7.4.2 Tourists' Central Concern When Traveling to the Galapagos Is Cost

Going to the Galapagos is expensive, easily costing US$3,000 to US$4,000 per person for international tourists (Epler 2007, 39; see Table 7.4 for further detail). According to travel agents, their customers do not want to pay for anything "extra"—they feel they are already paying enough. The general sentiment was that tourists cared most about getting "lo más barato [the cheapest]," "lo más económico [the most economic]," or a "good deal," instead of about being "green" or "sustainable." One tour operator-owner likened it to waiting for taxis in the rain: "You take the first one and you don't really care if it's a hybrid or not if it's pouring rain. You just want a cab. Ditto right now for the Galapagos ... people take the cheapest, most available trip. They're not checking for green." Betty, from Gala Mountain Travel, explained that tourists are beginning to ask a little bit more about sustainability, but usually only the highly educated—and it is

TABLE 7.4

Breakdown of Average Vacation Expenditures (US$) by Visiting Tourist Category

	Foreigners			Ecuadorians
Expenditure	Category 1	Category 2	Category 3	Category 4
International travel	1,320	1,448	1,911	NA
Other intl. expenditures*	548	240	NI	NA
In mainland Ecuador	874	666	617	NA
Air: Ecuador/Galapagos/ Ecuador	361	361	373	170
Galapagos cruise	2,454	1,538	1,062	55
Park fee and donations	139	110	110	7
In Galapagos towns	109	185	179	267
Hotels	20	82	43	92
Crafts	43	31	18	34
Meals	18	26	30	69
Other	28	46	88	72
Total	3,774	2,986	NI	478

Source: Bruce Epler, *Tourism, the Economy, Population Growth, and Conservation in Galapagos* (Puerto Ayora, Santa Cruz: Charles Darwin Foundation, 2007, p. 39. With permission.)
NA: Not applicable.
NI: Not included as expenditures listed; generally reflected only international air fares and expenditures in Ecuador and Galapagos.
* Expenditures in the country of residence and other countries visited.

still uncommon. Furthermore, some companies, like Happy Gringo for example, even offer cheap carbon offset programs (~US$10), but customers still say they will "think about it" and rarely purchase them.

While the tourists' focus on price (and not on sustainability/green) was corroborated by the passengers on board Ecoventura's vessel, "The Letty," they also highlighted additional reasons for selecting that particular tour operator—such as schedule flexibility, route and itinerary (which islands were visited), size of the boat, reviews from people who had traveled with Ecoventura in the past, informational website, and pictures. Only one couple (honeymooning and not budget-constrained) said that "green" and "responsible tourism" were their most important criteria. The woman had done a lot of research, felt Ecoventura was the greenest, and they were willing to pay even though "it was a little more expensive." They were in the minority. Another woman mentioned that she noticed the sustainability component while looking on Ecoventura's website (a project with wind turbines and Inuit villages) and that it "put them over." Although green and sustainability were not their main criteria, they were part of their decision-making process. Others, sometimes sheepishly, denied even having thought of sustainable/green, one woman claiming, "I'd love to say we picked it 'because it's green,' but really, it was just for the route."

7.4.3 Operators' Choice to "Be Sustainable" Still Boils Down to Profitability and Ethics

All but one of the tour operators contacted seemed to value sustainable tourism and described how they did their part to "be green." The company that did not value sustainable tourism could not describe anything it did that was sustainable and stated that its focus was, "Hacemos lo que quiere la gente [We do what the people want]." The other companies were proud to talk about their sustainable practices and why they had decided to focus on sustainability (out of their own personal choice and ethics, their company's ideology, etc.) even if the tourists were not yet demanding it or even asking about it. Such sustainability efforts included the following:

- Only working with tour operators in the Galapagos who run their *own* boats (as opposed to those who rent boats, who are less likely to actually be Galapagueños).
- Only working with people (hiring people) who are from the Galapagos.
- Using local guides, under the assumption that they also have a more vested interest in the land and animals; some also used different (local) guides for each island.
- Getting certified with "sellos verdes [green stamps]"—such as SmartVoyager, the Rainforest Alliance, etc. These certifications are *not* inexpensive: Smart Voyager costs US$2,500, and many of the certifications must be renewed or reapplied for annually.
- Using public transport between islands (instead of using their own boats).
- Asking tourists to re-use their plastic water bottles instead of using new ones each day.
- Teaching guides to cook differently (conserving more water and being more energy efficient).
- Using boats with solar power.
- Taking people to towns (thereby creating more jobs for locals and increasing the amount of money spent *in* the Galapagos).
- Offering carbon offset programs.
- Educating tourists and guides.

The facts that (1) sustainability can be expensive and requires additional effort, and that (2) it is still not demanded by tourists, leaves the decision to "be sustainable" (and likely sacrifice short-term profits for the long-term collective good) up to tour operators' ethics, values, and integrity. Although the tour operators and agencies advocated sustainability, it was still uncertain whether there is a great concern about it or whether the objective is to check the box to comply with legal standards. The *2007-2008 Galapagos Report* concluded, "An incentive mechanism must also be used to encourage the operators to move from mere compliance with an environmental 'checklist,' to maintaining their operations based on certified environmental management systems. This will ensure that the province's principal source of economic revenue be aligned with the fragile nature of the Galapagos" (Guime and Muñoz 2008, 72).

One could also see discrepancies between what was preached and what was practiced. For instance, although the Letty's captain and two head guides were supposed to be "from the Galapagos," all three of them actually lived on the mainland. Further, these three would be the leading wage earners on the vessel, and their money was returning to the mainland instead of staying on the islands—quite the reverse of ideal sustainability practices.

7.4.4 Ecuadorian (Domestic) Tourist Vision Misalignment

The literature clearly shows that Ecuadorian tourists (from the continent) stay on the Galapagos in hotels or with friends and family, which on the surface looks good for the local economy and monetary sustainability. But the travel agencies provided a better understanding of the Ecuadorian tourists' mindsets and reasons for going to the Galapagos—which are quite different from those of foreign tourists, and a bit alarming. One agency owner explained, "El turista nacional es lo peor [National tourists are the worst]," continuing on to say that they just want to go to the beach and enjoy themselves or have fun. Another shared that sentiment, explaining that the Galapagos are "expensive" and, therefore, the national tourists go to relax—not understanding that the Galapagos "es para entender [is for learning]" instead of for a relaxing summer beach break. The Galapagos "no es de lujo—es de conservar, de estar afuera [is not about luxury—it is about conservation and being outside]"—the visit should be about the people who live there, the animals, the natural setting, and the open air. The travel agencies further noted that foreign tourists were better educated about the Galapagos and had more positive mindsets that the Galapagos needed to be preserved and protected. The Ecuadorian tourists, on the other hand, were not "incentivados [motivated]" and should be more so. Again, there is an undeniable need for more education so that all tourists will feel similarly about the sustainability of the islands.

7.4.5 Tensions between Tour Operators and Retailers in the Galapagos Islands

Given the Galapagos residents' reliance on tourism, it is not surprising that the store owners on the Galapagos expressed a lot of negativity and frustration. They cannot control when tour operators bring tourists onshore or dictate how long they are allowed to stay on the island; yet their livelihood depends on the visitors' foot traffic. When tour operators bring tourists to see the giant tortoises and the Charles Darwin Foundation, in Puerto Ayora, Santa Cruz, the latter are normally given an hour or two to meander back to the port along the main street, with the expectation that they will go in and out of stores buying souvenirs. The store owners would like the tourists to spend more time on the islands because they believe that more time would bring in more money. Some also frustratingly pointed out that a lot of the resident islanders' money goes toward cleaning up trash and other items left behind by tourists and travel companies—two groups who are neither paying taxes on the islands nor spending much money on the islands.

When asked if the tour companies had done *anything* good for the islands, a few store owners were able to answer positively. According to storekeeper Pablo in Souvenir Pablis, one of the tour companies brings artists to show the locals new artistic techniques, to teach and train them. But he claimed that, in the end, it was still to the tour company's benefit because the artists usually worked for that agency. Another store worker, Estella from Edith II Pharmacy, discussed how tourism in general had helped the Galapagos community learn about recycling and why it is important. Islanders have learned to classify their trash (black = organic, green = inorganic, blue = recyclable), and there is no longer trash all over as there was when she was young. This is possibly less a function of the tour operators and more a function of the local island government trying to improve the community *for* tourism.

Several shopkeepers would prefer to see much more on-land tourism in order for the islanders to make more money and for tourism to be economically sustainable. One shopkeeper suggested changing the tourism model, gearing it more toward young people instead of older ones (as it is currently).

They could create more activities for the younger crowd, such as kayaking, and the youth would stay on the islands in hotels (because hotels are cheaper than live-aboard boats). However, one could still question whether more people, more hotels, and more activities on the inhabited islands would be economically beneficial only in the short term—and actually cause increasingly more harm to the environment and animals in the long run.

7.4.6 A Negative Cycle: Tourism Driving Migration, Driving Expansion, Driving Destruction

—"Todo hace daño [Everything does damage]." William Chica, Sales Manager, Ecuador View.

—"Siempre hay impacto [There is always an impact]." Christian Montesdeoca, General Manager, Ecomontes Tours.

Whereas some interviewed persons were adamant that tourism needs to be limited and/or reduced in the Galapagos, others talked more about how Galapagos natives should be allowed more licenses—and that more tourists should spend time on the islands.

If tourism quotas were reduced to protect the environment, yet migration to the islands continued to increase (driven by higher wages), the current island economic model would not be sustainable. Deporting residents (or illegal immigrants) and capping migration are possible courses of action—although the threat (and realization) of deportation angers current residents.

If tourism increased (or even remained the same), with an emphasis on lodging the tourists on the islands, then more infrastructure would need to be built to accommodate those tourists as well as the growing number of immigrants. The problem created with more infrastructure is, as Belisario Chiriboga, general manager of the Quito-based tour company, Tambopaxi, described, "that it will put more pressure on the flora and fauna." Similarly, if more tourism activities occur on the actual inhabited islands, such as kayaking and camping, while the islanders might earn more money (better for sustainable economics), the environment and animals will suffer. With enough destruction of the environment and animals, tourism would decrease, along with the residents' economic sustainability: a lose–lose for the environment, the communities, and the tourists. Given this negative cycle, there is no clear answer to the long-term sustainability of the Galapagos Islands other than better compliance incentives and/or increased enforcement of current laws and broader education.

7.4.7 Better Education and Stronger Law Enforcement

One common thread throughout almost all the interviews was the need for more education—for both foreign and national tourists, for tour operators, for guides, and, more broadly, for the entire Galapagos resident community. When asked what the most important changes could include, most shared the following sentiment: "Tenemos buenas leyes pero no las cumplen"—meaning there are good laws in place, but these laws are not well followed or enforced. Therefore, it is vital to continue "mejorando la educación [improving education]," especially targeting the youth. Various interviewees shared educational success stories from both the mainland and the Galapagos. They cited how, when schools taught children to recycle and not throw trash in the street, the children in turn questioned their parents if they did not recycle or if they disposed of waste indiscriminately—and how, over time, because of their children's persistence, parents learned and changed their behaviors. One of the Galapagueño storekeepers said he believed they needed political help because the economic system on the islands was based on political power and favors. He spoke more on the need for political reform than on the need for imposing new tourism laws—thereby also corroborating the tour operator's comments about the lack of follow-through on law enforcement. Paul Leon, from the Ministry of Tourism, also spoke to the importance of having tour operators involved with tourism policies at the governmental level (such as the Tourism Plan for 2020), so as to have their buy-in on their execution. He explained that if tour operators can be involved in creating such plans, then they have a stake in them and will benefit by following and implementing them. In addition, he believes that "instructing, educating, and involving the community" is also a requirement for success.

7.5 CONCLUSIONS

Is Galapagos tourism "sustainable"? The conclusions are not as rosy as one might wish. Although it is clear that people *want* the islands to be protected and preserved, how to do so economically and successfully is still in question. Institutions like the Charles Darwin Foundation and the World Wildlife Fund are undeniably creating a positive impact on the Galapagos' ecology and animals—through their research, preservation projects, hands-on work with animals, activism in legislation, and global awareness initiatives. However, the more overarching issue resides with the number of people coming (either tourists or immigrants) and how their mere presence is harmful to the sustainability of the archipelago. Although having more tourists on the islands does, indeed, bring more money and economic sustainability to residents, the requisite increase in infrastructure and land or water use damages the environment. Alternatively, if fewer tourists are allowed to come (which is more sustainable for the environment), then the resident population would need to be decreased.

Regardless of the tourism and immigration limitation concerns, education is a component of sustainability on which everyone does agree. Several proposed initiatives seem promising:

- Educating tour operators and guides about better sustainability practices.
- Educating Ecuadorian tourists about the experience they should expect on the islands when they come, thereby weeding out those who want only a beach vacation and teaching those who do come more about their own country and its uniqueness.
- Marketing/educating foreign tourists and philanthropists that the Galapagos are not "taken care of" and "protected" and that the "protection" is an ongoing process to which everyone must contribute.
- Educating tourists who visit the Galapagos about how to have minimal impact on the environment and the animals.
- Educating the resident communities in the Galapagos, especially the youth, about why their environment is so special, how best to protect it, and why this should be important to them.

Is tourism in the Galapagos actually sustainable? Much of the answer to this question will depend on whether tour operators, guides, and community members choose to maintain high ethical standards and persist in sustainable practices—at the expense of profit. If there are additional ways to entice Galapagos residents to return to the mainland (e.g., to go to a mainland university), this could also aid in making the Galapagos more sustainable. In short, the long-term preservation of the Galapagos seems to depend on each individual stakeholder incorporating into everyday actions the realization that sustainability practices must be followed. Better incentives toward compliance and enforcement (a top-down approach) together with increased education and communication (a bottom-up approach) will be crucial to the pursuit and attainment of the goal of sustainable tourism in the Galapagos Islands.

8

Conclusion
Green Product Innovation
and Adoption

João Neiva de Figueiredo and Mauro Guillén

CONTENTS

8.1 Systemic Approach 167
8.2 Incentive Alignment...168
8.3 Triple Bottom Line..169
8.4 Leading Globally and Acting Locally 170
8.5 Looking toward the Future 170

As exemplified by each of the cases in the previous chapters, the development and the adoption of green products and green energy alternatives have seen tremendous growth in recent years. Although human beings have shown remarkable capacity for adaptation and survival at the individual and at the community levels, the challenges we are expected to face in the centuries ahead suggest the need for concerted action at a broader scale. Population growth, climate change, social and income inequality, weapons of mass destruction, and public health failures are but a few examples of dangers that if left unchecked could threaten our very collective existence on the planet. Green technologies are expected to be an integral part of the antidotes to these threats. It is therefore advisable to understand not only successful patterns in the innovation and adoption of these technologies but also reasons for failure when that may have been the case.

Most green products imply some degree of behavioral change when introduced. Therefore it is necessary to examine the innovation and development of these products together with their adoption process. The cases of green production and consumption in this book attempt to do just that. They were selected with specific objectives in mind. First, they offer representative examples in different areas of focus within the green technology universe—they include energy provision (electricity), renewable fuel (ethanol), durable consumer products (automobiles), large-scale services (green urban mass transportation), and leisure infrastructure (tourism). Second, they offer perspectives from different parts of the world, including rich nations such as Denmark and developing nations such as Ecuador. Third, they represent very different categories of business activity, each with its own distinctive value chain. Finally, the cases rely on different institutional mechanisms to achieve widespread use, ranging from market incentives to extensive regulatory frameworks.

The cases described in this book are but a small sample illustrating the important role that the business sector has been called to play as we collectively build a more sustainable lifestyle on our planet. The United Nations' Bruntland Commission of 1987 defined sustainable development as "meeting the needs of the present without compromising the ability of future generations to meet their own needs" (United Nations General Assembly 1987). The term "sustainability" has since been expanded to aspects of human life that endeavor far beyond development and has come to represent the simultaneous fulfillment of environmental, social, and economic sustainability conditions. Broadly stated, environmental sustainability, also known as the "planet" component, relates to judicious use of natural resources, to pollution prevention, and to effective environmental management. Social sustainability, the "people" component, relates to equitable opportunity and wealth distribution, to fairness and justice, and to ethical behavior. Economic sustainability, the "profit" component, relates to the need for economic growth and the efficient use of resources.

Green product development and adoption is at the center of efforts toward increased global sustainability, which illustrates the central role businesses will play. It will be necessary for corporations and organizations to understand the hidden environmental and social costs of each product or service as well as each action taken. Recently there has been an effort toward pricing these hidden costs, but, because by their very nature, they are not amenable to market mechanisms, they are very difficult to quantify. Despite these complexities, business can tap the immense array of options to innovate and develop products that are truly sustainable in the sense of the Bruntland Commission's definition—that is, products that fulfill their objective without taxing future generations.

The cases described in this book are unique, but they are all examples of such efforts. The main messages from these cases seem to be not only that there is a need for orchestrated action at many levels, as will be described in the next section, but also that business will play a central role in the innovation and diffusion of products that lead to sustainable livelihoods. The disparity in the types of cases chosen to comprise this book enriches the exercise because it is apparent that, despite extreme differences, there are themes that are common to all the cases herein when comparing the processes of innovation, development, and adoption of the respective green technologies. All of these cases suggest that despite many victories, and despite the documented growth in green technologies, it is not easy to achieve large-scale changes in behavior patterns that lead to their widespread use. For green precepts to truly take hold in vastly different areas of economic production, such as agriculture, manufacturing, and services, businesses and consumers alike must have incentives respectively to provide and to select ever greener alternatives to existing products. It is not enough for a few conditions to be met if others are not.

Four common themes that are integral to each of the cases described in previous chapters are worth mentioning. First, the cases illustrate the need for a **systemic approach** to achieve successful adoption. Second, the cases indicate the advisability of **incentive alignment** among all stakeholders. Third, the cases stress the need for green technology solutions that are sustainable from the social, the environmental, and the economic points of view; that is, they must satisfy the **triple bottom line** of people, planet, and profit. Finally, the cases suggest the need for both a top-down and a bottom-up approach; that is, **leading globally and acting locally** is a necessary condition for successful adoption. Each of these four themes or tenets seems to be a necessary (although not sufficient) condition for successful widespread incorporation of green technologies to lifestyles throughout the world.

8.1 SYSTEMIC APPROACH

A common thread throughout the cases is the need for a systemic approach in which the different moving parts in the development and adoption of new technologies are addressed with attention to their many interrelationships. It is not enough to have a brilliant solution for one part of the puzzle if the linkages to related issues are not examined and solutions found. This is most apparent in the case of electricity where it is necessary to link the generation, transmission, and distribution components in a completely integrated system where sustainability objectives are met. Likewise, the development of electric cars that satisfy consumer needs and wants will not be completely successful without widespread implementation of a green plug-in refueling base, that is, large-scale public infrastructure to recharge automobile electric batteries. The bus rapid transit (BRT) system can only be successful as a green option when and if other related parameters, such as feeder systems and traffic patterns, are also accounted for.

The need for a systemic approach also applies to the sequencing of steps involved in the development and adoption of green technologies over time. Some innovations are incremental in that they do not involve drastic design changes to the product or to patterns of usage. Other innovations are revolutionary because they involve completely new designs of a product and/or patterns of utilization. Especially in the case of revolutionary changes, it is necessary to appropriately sequence the steps leading up to widespread adoption. Denmark might not have developed such a strong wind generation technology industry and therefore might not have reaped the full benefit of wind as a renewable energy source in the absence of the decentralized small producer combined heat power system. There is path dependency in successful green technology adoption.

8.2 INCENTIVE ALIGNMENT

Another common thread in the cases is the need for mechanisms that align the incentives of all stakeholders to ensure widespread adoption of a green technology. The first step in the process is necessarily to correctly identify the stakeholders and their objectives. This can be a complex process because often major stakeholders are not even at the table and therefore cannot explicitly defend their own point of view. Use of a BRT system is not enough on the one hand without accompanying incentives to discourage driving by car owners and on the other hand without help for underprivileged residents to use and take most advantage of the system. In the Galapagos Islands it is necessary to provide incentives not only for the tour operators and the tourists but also for the local population, including disincentives to immigration. The Danish model of alternative renewable energy sourcing flourished partly as a result of incentive systems that addressed the needs of all major stakeholders, including consumers, researchers, industry, and investors.

Many green initiatives can rely solely on market mechanisms such as consumer demand to address some stakeholder incentive issues, but most require some sort of additional incentive somewhere in their development or adoption processes. The need to move down the learning curve in order to achieve the economies necessary for lower cost production and consequent affordable pricing vis-à-vis substitute products may require incentive mechanisms for producers in developmental stages of a new technology. The electric engines are a case in point. Without broad-based institutional and governmental support in the face of resistance by opposing interests, in the late 1990s GM shelved a promising electric car initiative before it became fully operational.

8.3 TRIPLE BOTTOM LINE

Much has been written about the concept of the triple bottom line. It is used here solely as a "catch-all" term that can easily be understood and that expresses the desirability of integrating social, environmental, and economic considerations in any large-scale green alternative solution. The need for economic sustainability is clearly understood—in fact it has been the object of study at business schools for over one hundred years. What has become clear in recent decades is the absolute need to incorporate environmental and social sustainability considerations in any endeavor.

It is not enough to examine tourism in the Galapagos with an eye to environmental sustainability without analyzing the social impact on the local population; that is, changes in patterns of tourism must be socially sustainable as well. The sugarcane ethanol option as a renewable automotive fuel energy source will only be complete when the social issues of sugarcane cutters are addressed in a way that is long-term sustainable. As highlighted in the electricity generation case, over the past thirty years, Denmark has been able to lower energy consumption while achieving GDP growth and addressing social and environmental issues, undoubtedly a very encouraging outcome. The ongoing public-private partnerships in the Mexico City BRT project, in which the absence of subsidies are noteworthy, are yet another successful example of the need for a three-pronged approach focusing on economic, environmental, and social sustainability.

8.4 LEADING GLOBALLY AND ACTING LOCALLY

The green production and consumption movement includes the development of new green technologies to be sure, but it also includes an evolution in consumer behavior. One sign of this evolution is the growing perception that each consumer is responsible for the product after usage, that is, that waste management is ultimately the responsibility of each and every one of us. Another sign is the increasing impact of grass-roots movements in tackling sustainability issues. In order to reach full adoption, green technologies need both unambiguous top-down direction and inclusive bottom-up action. One small-scale example of this multiplicative effect is the positive impact of careful and informed tourists in an environment with a clear top-down mandate favoring sustainability as described in the Galapagos case. There are large-scale examples as well. Denmark's continuous ability to pair sustainable government policies, regulations, and initiatives with constructive response at the local level has been instrumental throughout the country's successful development and adoption of alternative energy sources. Likewise, Brazilian government policies stimulating ethanol fuel production and distribution were matched by a consumer base that was willing to try this alternative renewable automotive fuel option, buying cars that ran only on ethanol in the early stages of the program, and providing the initial critical mass that allowed for the ensuing development and growth of a sophisticated industry.

8.5 LOOKING TOWARD THE FUTURE

The universe of green products is a huge landscape that encompasses virtually every product and service category and which has the potential of providing a needed foundation to a truly sustainable society. This book began with a broad map of the green product universe, including energy sources; energy use; environmental and recyclable products and materials; water management; green consumption; implications for governments, including policies and regulation; and implications for businesses, including effects on different parts of the value chain. This map was offered with the objectives of positioning the cases described in the book and of illustrating the many different possible avenues for positive impact from green products. The cases described in the book are examples that offer insights into the systemic approach that is required in order to create a truly greener world. We hope that lessons learned from these cases may be applied in other situations going forward.

Bibliography

Aalbers, R., van der Heijden, E., Potters, J., van Soest, D., and Vollebergh, H. 2009. Technology Adoption Subsidies: An Experiment with Managers. *Energy Economics* 31:431–442.

Åhman, M. 2006. Government Policy and the Development of Electrical Vehicles in Japan. *Energy Policy* 34:433–443.

Albino V., Balice, A., and Dangelico, R.M. 2009. Environmental Strategies and Green Product Development: An Overview on Sustainability-Driven Companies. *Business Strategy and the Environment* 18:83–96.

Alvarado, P. 2007. Mexico Approves Corn and Sugar Cane Ethanol Law. *Treehugger: A Discovery Company.* April 30. http://treehugger.com/files/2007/04/mexico_approves. php (accessed September 2009).

ANFAVEA (Associação Nacional dos Fabricantes de Veículos Automotores, Brasil). 2009. *Produção de automóveis por tipo e combustível, 2009* (Tabela 10). São Paulo: ANFAVEA. http://www.anfavea.com.br/tabelas/autoveiculos/tabela10_producao. pdf (accessed July 17, 2009).

ASTM Standard D6866-06. 2009. *Standard Test Methods for Determining the Bio-based Content of Natural Range Materials Using Radiocarbon and Isotope Ratio Mass Spectrometry Analysis.* ASTM International, West Conshohocken, PA. www.astm.org (accessed November 2009).

Atkinson, R. 2008. Advances in the Conservation of Threatened Plant Species of Galapagos. In *Galapagos Report 2007-2008.* Puerto Ayora, Galapagos, Ecuador: CDF, GNP, & INGALA.

Atkinson, R., Rentería, J.L., and Simbaña, W. 2008. The Consequences of Herbivore Eradication on Santiago: Are We in Time to Prevent Ecosystem Degradation Again? In *Galapagos Report 2007-2008.* Puerto Ayora, Galapagos, Ecuador: CDF, GNP, & INGALA.

Ayres, R.U., Turton, H., and Casten, T. 2007. Energy Efficiency, Sustainability and Economic Growth. *Energy* 32:634–648.

Baumann H., Boons, F., and Bragd, A. 2002. Mapping the Green Product Development Field: Engineering, Policy and Business Perspectives. *Journal of Cleaner Production* 10:409–425.

Birol, F. 2004. *Analysis of the Impact of High Oil Prices on the Global Economy.* International Energy Agency Report. http://www.iea.org/work/2004/cambodia/bj_session1.3.pdf

Bonini, S.M., and Oppenheim, J.M. 2008. Helping "Green" Products Grow. *McKinsey Quarterly* (2008):1–8.

Book, M., Groll, M., Mosquet, X., Rizoulis, D., and Sticher, G. 2009. *The Comeback of the Electric Car?* Boston Consulting Group (BCG) Report. http://www.bcg.com/docu-ments/file15404.pdf (accessed May 19, 2010).

Boon, M. 2008. *Why Did Danish Entrepreneurs Take the Lead in the Wind Turbine Industry and Not the Dutch?* MSc Thesis, RSM Erasmus University.

Bourne, J.K., Jr. 2007. Green Dreams. *National Geographic Magazine* 212(4):38–59.

Breakthrough Institute. 2009. Inheriting the Wind—Danish Wind Power. In *Case Studies in American Innovation: A New Look at Government Involvement in Technological Development*. Oakland, CA: Breakthrough Institute.

Brown, L.R. 2006. Beyond the Oil Peak. In *Plan B 2.0: Rescuing a Planet under Stress and a Civilization in Trouble*. New York: Norton.

Brueckner, J.K. 2000. Urban Sprawl: Diagnosis and Remedies. *International Regional Science Review* April:160–171.

Bryce, R. 2008. The Corn Ethanol Juggernaut. *Environment 360*, September 15. http://e360. yale.edu/feature/the_corn_ethanol_juggernaut/2063 (accessed January 28, 2010).

Budeanu, A. 2005. Introduction and Overview of the *Journal of Cleaner Production* Special Issue on Sustainable Tourism. *Journal of Cleaner Production* 13:79–81.

Burkhard, R., Deletic, A., and Craig, A. 2000. Techniques for Water and Wastewater Management: A Review of Techniques and Their Integration in Planning. *Urban Water* 2:197–221.

Buzzanell, P.J. 1992. Cuba's Sugar Industry—Facing a New World Order. *Situation and Outlook Report: Sugar and Sweetener*, SSRV17N1. Washington, DC: Economic Research Service, U.S. Department of Agriculture, March, pp. 24–41.

Cantú, R. 2007. *Ethanolomics: The Think-about's of the Mexican Ethanol Project*. Working Paper No. 2007-3, Cátedra de Integración Económica y Desarrollo Social, Escuela de Graduados en Administración Pública y Política Pública, Technológico de Monterrey. August. http://www.mty.itesm.mx/egap/deptos/cee/cieds/2007-3.pdf

CARB (California Air Resources Board). 2009. *California State Zero-Emission Mandate*. http://www.arb.ca.gov/msprog/zevprog/zevprog.htm (accessed April 10, 2010).

CDF, GNP, & INGALA. 2008. *Galapagos Report 2007-2008*. Puerto Ayora, Galapagos, Ecuador: CDF, GNP, & INGALA.

Ceballos-Lascurain, H. 1991. Tourism, Eco-tourism and Protected Areas. In Kusler, J.A. (Ed.), *Eco-tourism and Resource Conservation, vol. 1, Eco-tourism and Resource Conservation Project*. Waterloo, Ontario: Heritage Resources Centre, University of Waterloo.

Central Intelligence Agency. 2010. Denmark. In *The World Factbook*. April 1. https://www.cia.gov/library/publications/the-world-factbook/geos/da.html (accessed April 2010).

Choi, H.C., and Sirakaya, E. 2006. Sustainability Indicators for Managing Community Tourism. *Tourism Management* 27:1274–1289.

Christensen, J., Dalum, B., Gregersen, B. Johnson, B., Lundvall, B.-Å., and Tomlinson, M. 2005. *The Danish Innovation System*. Paper presented at the Seoul Workshop. March 7–9.

Christianson, R. 2005. Danish Wind Co-ops Can Show Us the Way. *Wind Works*. http://www.wind-works.org/articles/Russ%20Christianson%20NOW%20Article% 201.pdf (accessed April 2010).

Chung, Y., and Tsai, C. 2007. The Effect of Green Design Activities on New Product Strategies and Performance: An Empirical Study among High-tech Companies. *International Journal of Management* 24:276–288.

Cohen, B. 2004. Urban Growth in Developing Countries: A Review of Current Trends and a Caution Regarding Existing Forecasts. *World Development* 32(1):23–51.

Cohen, J. 2001. *World Population in 2050: Assessing the Projections*. Paper presented at the Federal Reserve Bank of Boston's Annual Research Conference Series.

Confederation of Danish Industry – DI. 2010. 10 Facts about the Danish Climate Technology Business. In *Solutions for Sustainable Growth*. Copenhagen: Confederation of Danish Industry – DI, pp. 10–11.

Constance, P. 2007. The Age of Ethanol? *IADB Magazine of the Inter-American Development Bank.* December.

Cubero-Pardo, P., and Bastidas, E.A. 2008. The Impact of Ecotourism Activities on Wildlife and Sessile Benthic Species in the Galapagos Marine Reserve. In *Galapagos Report 2007-2008.* Puerto Ayora, Galapagos, Ecuador: CDF, GNP, & INGALA.

Danish Energy Agency. 2008. *Energy Statistics 2008.* Copenhagen: Danish Energy Agency.

Danish Energy Agency. 2009. *Denmark: A Leading Player in Combined Heat and Power.* Copenhagen: Danish Energy Agency. http://www.ens.dk/en-US/Info/news/Factsheet/Documents/kraftvarme%20170709.pdf

Danish Energy Agency. n.d. Public Heat Planning (1970s and 1980s). http://ens.dk/en-US/supply/Heat/Goals_and_means/planning/Sider/Forside.aspx (accessed April 2010).

Danish Energy Authority. 2005a. *Energy Strategy 2025: Perspectives to 2025 and Draft Action Plan for the Future Electricity Infrastructure.* Copenhagen: Danish Ministry of Transport and Energy.

Danish Energy Authority. 2005b. *Heat Supply in Denmark: Who What Where and Why.* Copenhagen: Danish Energy Authority.

Darwin, C. 1859. Geographical Distribution. In *Origin of the Species.* http://www.literature.org/authors/darwin-charles/the-origin-of-species/chapter-12.html (accessed November 5, 2009).

Davis, S.C., and Diegel, S.W. 2005. *Transportation Energy Data Book: Edition 24.* Center for Transportation Analysis, Oak Ridge National Laboratory. http://www.ornl.gov/~webworks/cppr/y2001/rpt/122271.pdf (accessed March 7, 2011).

de Lima, J.G. 2006. A Riqueza é o Saber. *Revista Veja.* January 1. http://veja.abril.com.br/010206/p_096.html (accessed July 14, 2009).

Dean, A. 2007. Unethical Ethanol Tariff. *Policy Innovations.* April 4. http://www.policyinnovations.org/ideas/briefings/data/ethanol (accessed September 2009).

Dzioubinski, O., and Chipman, R. 1999. *Trends in Consumption and Production: Household Energy Consumption.* DESA Discussion Paper No. 6.

Earth Policy Institute. 2010. Supporting data for Chapters 4 and 5—Geothermal. In Brown, L.R. (ed.), *Plan B 4.0: Mobilizing to Save Civilization.* New York: Norton. http://www.earth-policy.org/datacenter/pdf/book_pb4_ch4-5_geo_pdf.pdf (accessed July 18, 2010).

Ecuadorian Ministry of Tourism. 2007. *Informe Ejecutivo: Plan Estratégico de Desarrollo de Turismo Sostenible para Ecuador PLANDETUR 2020.* Quito: Ecuadorian Ministry of Tourism.

Electrification Coalition. 2009. *Electrification Roadmap: Revolutionizing Transportation and Achieving Energy Security.* November. Washington, DC: Electrification Coalition.

EMBARQ and the Center for Sustainable Transport in Mexico. 2006a. Bus Rapid Transit: Arriving on the World Scene. In *Sustainable Mobility,* 16–21. http://www.embarq.info/sites/default/files/Metrobus_Welcome_Aboard.pdf

EMBARQ and the Center for Sustainable Transport in Mexico. 2006b. Crossing Mexico City: A Vale of Tears. In *Sustainable Mobility,* 30–32. http://www.embarq.info/sites/default/files/Metrobus_Welcome_Aboard.pdf

EMBARQ and the Center for Sustainable Transport in Mexico. 2006c. For Cleaner Air. In *Sustainable Mobility,* 33–36. http://www.embarq.info/sites/default/files/Metrobus_Welcome_Aboard.pdf

EMBARQ. 2009a. *Macrobús Scores High in User Acceptance.* May 15. Washington, DC: World Resources Institute. http://www.embarq.org/en/news/09/05/15/macrob%C3%BAs-scores-high-user-acceptance

EMBARQ. 2009b. *Mexico City.* Washington, DC: World Resources Institute. http://www. embarq.org/en/city/mexico-city-mexico (accessed December 11, 2009).

EMBARQ. 2009c. *Mexico City, Metrobús: A Project That's Changed the City.* http://www. embarq.org/en/project/mexico-city-Metrobus

Empresa de Pesquisa Energética. 2008. *Balanço energético nacional 2008: Ano base 2007.* Ministério de Minas e Energia do Brasil. Tables 3.6a and 3.6b.

Energy Information Administration (EIA). 2007. *Electric Power Annual 2007—State Data Tables.* http://www.eia.doe.gov/cneaf/electricity/epa/epa_sprdshts.html (accessed April 10, 2010).

Epler, B. 2007. *Tourism, the Economy, Population Growth, and Conservation in Galapagos.* Puerto Ayora, Santa Cruz, Ecuador: Charles Darwin Foundation.

European Commission. 2008. *EU Energy Security and Solidarity Action Plan: 2nd Strategic Energy Review Summary.* http://www.europa-eu-un.org/articles/en/article_8300_en.htm

European Renewable Energy Council. 2004. *Renewable Energy Scenario to 2040.* Brussels: European Renewable Energy Council.

Eurostat, European Commission. 2008. *Energy Production and Imports—Statistics Explained.* September. http://epp.eurostat.ec.europa.eu/statistics_explained/index. php/ Energy_production_ and_imports (accessed April 2010).

Field, C.B., Campbell, J.E., and Lobell, D.B. 2008. Biomass Energy: The Scale of the Potential Resource. *Trends in Ecology and Evolution* 23:65–72.

Fischer, G., Teixeira, E. Tothne Hizsnyik, E., and van Velthuizen, H. 2008. Land Use Dynamics and Sugarcane Production. In *Sugarcane Ethanol: Contributions to Climate Change Mitigation and the Environment*, P. Zuurbier and van de Vooren, J., Eds. 29–62. Wageningen, The Netherlands: Wegeningen Academic. http://www.globalbioenergy. org/uploads/media/ 0811_Wageningen_-_Sugarcane_ethanol__Contributions_to_ climate_change_mitigation_and_the_environment.pdf

Fridleifsson, I.B. 1996. Present Status and Potential Role of Geothermal Energy in the World. *Renewable Energy* 8:34–39.

Fthenakis, V., Mason, J.E., and Zweibel, K. 2009. The Technical, Geographical, and Economic Feasibility for Solar Energy to Supply the Energy Needs of the U.S. *Energy Policy* 37:387–399.

Galapagos National Park. 2009. Control de la operación turística. http://www.galapago-spark.org/programas/turismo_control.html (accessed November 5, 2009).

Gallup, D.L. 2009. Production Engineering in Geothermal Technology: A Review. *Geothermics* 38:326–334.

Giusti, L. 2009. A Review of Waste Management Practices and Their Impact on Human Health. *Waste Management* 29:2227–2239.

Glasnovic, Z., and Margeta, J. 2009. The Features of Sustainable Solar Hydroelectric Power Plant. *Renewable Energy* 34:1742–1751.

Glavic, P., and Lukman, R. 2007. Review of Sustainability Terms and Their Definitions. *Journal of Cleaner Production* 15:1875–1885.

Global Waste Management Market Assessment 2007. Key Note. http://www.keynote .co.uk/market-intelligence/view/product/1903/global-waste-management? highlight=waste&utm_source=kn.reports.search (accessed March 10, 2011).

Gogate, P.R., and Pandit, A.B. 2004. A Review of Imperative Technologies for Wastewater Treatment. *Advances in Environmental Research* 8:501–597.

Goldemberg, J. 2008. *Biotechnology for Biofuels.* São Paulo: University of São Paulo, Institute of Electrotechnics and Energy.

Goldemberg, J., Coelho, S.T., and Guardabassi, P. 2008. The Sustainability of Ethanol Production from Sugarcane. *Energy Policy* 36:2086–2097.

Gómez Macías, I. 2008. *Mexico's Entrance into the Bioenergetics World.* Paper presented at the GEC Meeting, Chicago, April. http://www.governorsbiofuelscoalition.org/assets/files/Meeting%20Presentations/Mexico-Isabel-4-08.pdf (accessed October 1, 2009).

Graham, C. 2006. Bus Rapid Transit in Australasia: Performance, Lessons Learned and Futures. *Journal of Public Transportation, Special Edition: BRT* 9:1–22.

Graves, C., ed. 2006. *Charles Darwin Foundation Strategic Plan 2006-2016.* Puerto Ayora, Santa Cruz, Ecuador: Charles Darwin Foundation.

Green Car Congress. 2007. *São Paulo Puts Ethanol Bus into Service in BEST Project.* December 23. http://www.greencarcongress.com/2007/12/so-paolo-puts-e.html (accessed March 14, 2008).

Guime, M.P., and Muñoz, E.H. 2008. General Characteristics of the Tourist Fleet in Galapagos and Its Compliance with Environmental Standards. In *Galapagos Report 2007-2008.* Puerto Ayora, Galapagos, Ecuador: CDF, GNP, & INGALA.

GWEC. 2009. *Global Wind 2009 Report.* Brussels: Global Wind Energy Council (GWEC).

Gwilliam, K., Kojima, M., and Johnson, T. 2004. *Reducing Air Pollution from Urban Transport.* Washington, DC: World Bank.

Halkier, B. 2001. Risk and Food: Environmental Concerns and Consumer Practices. *International Journal of Food Science and Technology* 36:801–812.

Hammar, T. 1999. *The Case of CHP in Denmark and Perspectives to Other Countries.* Input to the Annex I Workshop on Energy Supply Side Policies and Measures. September 10.

Hansen, M.D. 2001. *The Danish Experience with Efficiency Improvement in Industrial and Commercial Sectors.* Paper presented at Workshop on Best Practices in Policies and Measures in Copenhagen. October 8–10.

Havas Media. 2008. *Consumer Perception of Climate Change and Its Potential Impact on Business.* Executive Summary. http://www.havasmedia.com/staticfiles/executive_summary.pdf (accessed April 2010).

Herring, H. 2006. Energy Efficiency: A Critical View. *Energy* 31:10–20.

Heyes, A. 2000. Implementing Environmental Regulation: Enforcement and Compliance. *Journal of Regulatory Economics* 17:107–129.

Hidalgo, D. 2006. *High Level BRT: An Option to Consider Even at Very High Demand Levels.* Paper presented at the 5th International Bus Conference, Bogotá. February 14.

Hofstrand, D. 2009. Brazil's Ethanol Industry. *AgDM Newsletter.* January.

Honda. 2008. *Honda Insight Concept Hybrid Vehicle to Debut at Paris International Auto Show.* Honda Corporate Press Release. September 14. http://www.hondanews.com/categories/939/releases/4722 (accessed April 10, 2010).

Hopkins, S. 2009. Petrobras Keeps Brazil Ahead in Bio-Fuel. *Matter Network New Ideas for a Sustainable World.* March 13. http://www.matternetwork.com/2009/3/petrobras-keeps-brazil-ahead-biofuel.cfm (accessed September 2009).

House of Representatives. 2009. HR2454, American Clean Energy and Security Act of 2009, June 26. http://thomas.loc.gov/cgi-bin/bdquery/z?d111:HR02454:@@@D&s u mm2=m&

Husain, I. 2003. *Electric and Hybrid Vehicles. Design Fundamentals.* Boca Raton, FL: Taylor & Francis CRC Press.

Hvelplund, F. 2006. *Political Liberalization and Green Innovation at the Danish Energy Scene.* Aalborg: Aalborg University.

INGALA. 2009. http://www.ingala.gov.ec (accessed November 5, 2009).

INGALA. 2011. Galapagos Plants. The Galapagos National Institute (INGALA). http://
www.ingala.gob.ec/galapagosislands/index.php?option=com_content&task=view&i
d=71&Itemid=69 (accessed March 15, 2011).

Inman, R.E. 2009. *Making Cities Work. Prospects and Policies for Urban America.* Princeton,
NJ: Princeton University Press.

International Energy Agency (IEA). 2008a. *Denmark—Answer to a Burning Platform: CHP/
DHC.* http://www.iea.org/g8/chp/docs/denmark.pdf (accessed April 2010).

International Energy Agency (IEA). 2008b. *World Energy Outlook 2008.* OECD/IEA.

International Energy Agency (IEA). 2009. *World Energy Outlook 2009.* OECD/IEA.

International Energy Agency (IEA). 2010. *CHP/DHC Country Scorecard: Denmark.* The
International CHP/DHC Collaborative. http://www.iea.org/G8/CHP/profiles/den-
mark.pdf (accessed April 2010).

IPCC. 2007. Summary for Policymakers. In *Climate Change 2007: Mitigation. Contribution
of Working Group III to the Fourth Assessment Report of the Intergovernmental Panel
on Climate Change.* Cambridge: Cambridge University Press.

Irrisoft. 2009. *Weather Reach—Landscape Water Management.* North Logan, UT:
Irrisoft, Inc. http://www.irrisoft.net/downloads/manuals/Landscape%20Water%20
Management%20Training%20Manual.pdf (accessed November 3, 2009).

ISRI. 2009. Design for Recycling: The Future is Now. Institute of Scrap Recycling
Industries, Inc. (ISRI). http://www.isri.org/Content/NavigationMenu/Advocacy/
DesignForRecycling/default.htm (accessed November 1, 2009).

Jackson, P. 2009. *The Future of Global Oil Supply: Understanding the Building Blocks.*
Cambridge: IHS CERA.

Jacobs, J. 2006. Ethanol from Sugar: What are the Prospects for U.S. Sugar Co-ops? *USDA
Rural Development.* October. http://findarticles.com/p/articles/mi_m0KFU/is_5_73/
ai_n27014218

Janssen, R. 2008. *Energy Efficiency Policy Explained: An Introduction.* March.
www.helio-international.org/EEPolicyExplained.pdf

Jasch, C. 2006. Environmental Management Accounting (EMA) as the Next Step in the
Evolution of Management Accounting. *Journal of Cleaner Production* 14:1190–1193.

Jensen, C. 2009. Ethanol Industry's 15% Solution Raises Concerns. *New York Times.* May 8.

John, G., Clements-Croome, D., and Jeronimidis, G. 2005. Sustainable Building Solutions:
A Review of Lessons from the Natural World. *Building and Environment* 40: 319–328.

Jones, M.T. 1996. Social Responsibility and the "Green" Business Firm. *Industrial and
Environmental Crisis Quarterly* 9:327–345.

Joshi, S., Krishman, R., and Lave, L. 2006. Estimating the Hidden Costs of Environmental
Regulation. *Accounting Review* 76:171–198.

Kale, G., Kijchavengkul, T., Auras, R., Rubino, M., Selke, S., and Singh, S.P. 2007.
Compostability of Bioplastic Packaging Materials: An Overview. *Macromolecular
Bioscience* 7:255–277.

Kalogirou, S.A. 2004. Environmental Benefits of Domestic Solar Energy Systems. *Energy
Conversion and Management* 45:3075–3092.

Karki, R. 2007. Renewable Energy Credit Driven Wind Power Growth for System Reliability.
Electric Power Systems Research 77:797–803.

Kawai, T., and Niikuni, T. 2009. Characteristics of Plug-In Hybrid Electric Vehicle and
Problems for Performance Assessment. *Journal of the Society of Automotive Engineers*
63(9):29–36 (In Japanese).

Kerin, P.D. 1992. Efficient Bus Fares. *Transport Reviews* 12:33–47.

Khan, S., Tariq, R., Yuanlai, C., and Blackwell, J. 2006. Can Irrigation Be Sustainable? *Agricultural Water Management* 80:87–99.

Khawaji, A.D., Kutubkhanah, I.K., and Wie, J.-M. 2008. Advances in Seawater Desalination Technologies. *Desalination* 221:47–69.

Kjems, J. 2009. Clean Energy—Danish Style. *Living Energy*, November.

Larminie, J., and Lowry, J. 2003. *Electric Vehicle Technology Explained*. London: John Wiley & Sons.

Leonard, A. 2007. *How the World Works: The Seven Commandments of Mexican Ethanol*. http://www.salon.com/tech/htww/2007/09/28/mexican_ethanol_economics (accessed September 29, 2009).

Licht, F.O. 2008. Brazil's Ethanol Exports to Rise in 2008. *Reuters*. March 5. http://uk.reuters.com/article/oilRpt/idUKB53020120080305?pageNumber=1&virtualBrandChannel=0 (accessed September 29, 2009).

Lior, N. 2008. Energy Resources and Use: The Present Situation and Possible Paths to the Future. *Energy* 33:842–857.

Litman, T. 2005. Efficient Vehicles versus Efficient Transportation. Comparing Transportation Energy Conservation Strategies. *Transport Policy* 12:121–129.

Loe, D.L. 2003. Quantifying Lighting Energy Efficiency: A Discussion Document. *Lighting Research & Technology* 35(4):319–329.

Lombrano, A. 2009. Cost-efficiency in the Management of Solid Urban Waste. *Resources, Conservation and Recycling* 53:601–611.

Luhnow, D., and Samor, G. 2006. As Brazil Fills up on Ethanol, It Weans off Energy Imports. *Wall Street Journal*. January 16.

Lund, H. 1999. Implementation of Energy-Conservation Policies: The Case of Electric Heating Conversion in Denmark. *Applied Energy* (September 1):112–117.

Mackey, M. 2009. Colombia Ethanol Aided by Government Regulations: Support for New Sector Consolidates Reformers in South America. *Suite101.com Insightful Writers. Informed Readers*. April 24. http://colombia.suite101.com/article.cfm/colombia_ethanol_aided_by_government_regulations (accessed September 2009).

Madding, A. 2008. NASCAR—Why Not Ethanol? *Insider Racing News.com*. June 24. http://insiderracingnews.com/Writers/AM/062408.html (accessed September 29, 2009).

Maegaard, P. 2009. Danish Renewable Energy Policy. *World Council for Renewable Energy*. http://www.wcre.de/en/images/stories/pdf/WCRE_Maegaard_Danish%20RE%20Policy.pdf

Mainichi Shimbun. 2009. Prius Wins New Vehicle Sales for 6 Months in a Row. *Mainichi Shinbum*. April 12. http://mainichi.jp/select/today/news/20091204k0000e020066000c.html (accessed April 10, 2010).

Mason, V.D. 2008. *Wind Power in Denmark*. November. http://www.wind-watch.org/documents/wp-content/uploads/mason-windpowerindenmark-2008.pdf

Mayor of London. 2009. *An Electric Vehicle Delivery Plan for London*. May. http://legacy.london.gov.uk/mayor/publications/2009/docs/electric-vehicles-plan.pdf (accessed May 19, 2010).

McKendry, P. 2002. Energy Production from Biomass (part 1): Overview of Biomass. *Bioresource Technology* 83:37–46.

McKinsey. 2007. *How Companies Think About Climate Change: A McKinsey Global Survey*. December. http://www.mckinsey.com/clientservice/sustainability/pdf/climate_change_survey.pdf (accessed April 2010).

MEE. 2006. *Alianza entre Ecopetrol y Petrobrás para desarrollo de Biocombustibles*. Press Release, Ministry of Mines and Energy, Colombia. October 17.

MEJ (Ministry of Environment of Japan). 2008. Ministry White Paper. http://www.env. go.jp/policy/hakusyo/h20/index.html (accessed April 10, 2010).

Metge, H., and Jehanno, A. 2006. *A Panorama of Urban Mobility Strategies in Developing Countries*. Washington, DC: World Bank.

Meyer, N. 2007. Learning from Wind Energy Policy in the EU: Lessons from Denmark, Sweden and Spain. *European Environment* (September 27): 347–362.

Ministry of Climate and Energy of Denmark. 2009. *"The Danish Example": The Way to an Energy Efficient and Energy Friendly Economy*. Copenhagen: Ministry of Climate and Energy.

Ministry of Foreign Affairs of Denmark. 2008. A Nation Coping with Climate Change. *Mandag Morgen Special Edition*, November.

Mintel. 2009. *Green Living—US*. Research Report. Chicago: Mintel Oxygen.

Mitric, S. 2008. *Urban Transport for Development towards an Operationally Oriented Society*. Washington, DC.

Mitsubishi. 2009. *Mitsubishi Motors Lineup at 2009 Geneva International Motor Show*. Press Release. March 3. http://www.mitsubishi-motors.com/publish/pressrelease_en/ motorshow/2009/news/detail1906.html (accessed April 10, 2010).

Mitsubishi. 2010. *Lithium Energy Japan Commits to Construction of New Plant: New Ritto Plant (Japan) to Begin Full Mass Production of Lithium-ion Batteries for 50,000 Electric Vehicles per Year in 2012*. Press Release. April 14. http://www.mitsubishi-motors. com/publish/pressrelease_en/index.html

Mochimaru, T., and Lee, E. 2009. Electric-Car Market Starting to Move. Tokyo: Barclays Capital Report.

Montalvo, C. 2008. General Wisdom Concerning the Factors Affecting the Adoption of Cleaner Technologies: A Survey 1990-2007. *Journal of Cleaner Production* 16:S7–S13.

Montalvo, C., and Kemp, R. 2008. Cleaner Technology Diffusion: Case Studies, Modeling and Policy. *Journal of Cleaner Production* (editorial) 16:S1–S6.

Nikkan Kogyo Shimbun. 2009. *Daily Industrial News*. July 20, p. 22.

Nikkei Shimbun. 2009a. December 2. http://www.nikkei.co.jp/news/sangyo/20091203AT1 C0200E02122009.html (accessed March 1, 2010).

Nikkei Shimbun. 2009b. May 19, p. 3.

Nikkei Trendy Net. 2009. Nikkei Trendy Net. March 8. http://trendy.nikkeibp.co.jp/article/ pickup/20090803/1028073/?P=1 (accessed April 10, 2010).

Nissan. 2010a. Nissan Leaf promotion website, http://www.nissanusa.com/leaf-electric-car (accessed April 10, 2010).

Nissan. 2010b. *Nissan Will Launch New Season for Mobility in April by Starting Pre-orders for Nissan LEAF*. Press Release. March 30. http://www.nissan-global.com/EN/ NEWS/2010/_STORY/100330-01-e.html (accessed April 10, 2010).

Nowosielski, R., and Zajdel, A. 2007. Recycling's technology. *Journal of Achievements in Materials and Manufacturing Engineering* 21:85–88.

Obama, B. 2009. Opening Remarks, The Fifth Summit of the Americas, Port of Spain, Trinidad and Tobago, April 2009. http://www.fifthsummitoftheamericas.org/official_ statements(2).htm (accessed October 5, 2009).

OECD. 1999. *Denmark—Regulatory Reform in Electricity*. Paris: OECD.

OECD. 2005. *Opening Markets for Environmental Goods and Services*. Paris: OECD Observer Policy Briefs.

OECD. 2008. *OECD Environmental Outlook to 2030*. Paris: OECD.

OECD. 2010. *Index of Statistical Variables. January 2010*. http://www.oecd.org/ dataoecd/26/40/38785295.htm (accessed April 2010).

Olesen, G.B. n.d. *Danish Initiatives and Planning in the Field of Energy Efficiency and Renewable Energy.* International Network for Sustainable Energy. http://www.inforse. dk/europe/word_docs/s_gbo_dk.doc (accessed April 2010).

Olsson, L.E., and Gärling, T. 2008. Staying Competitive While Subsidized: A Governmental Policy to Reduce Production of Environmentally Harmful Products. *Environment and Planning C: Government and Policy* 26:667–677.

Paine, C. 2006. *Who Killed the Electric Car?* November 14. Sony Pictures Home Entertainment.

Parry, I.W.H., and Small, K.A. 2009. Should Urban Transit Subsidies Be Reduced? *American Economic Review*: 700–724.

Patel, T. 2009. *Immigration Issues in the Galapagos Islands.* Washington, DC: Council in Hemispheric Affairs. http://www.coha.org/immigration-issues-in-the-galapagos-islands (accessed November 5, 2009).

Pedersen, C.D. 2005. *EU Renewable Energy Policy: A Study on the EU Legislative Framework for Promoting Renewable Energy.* May. Roskilde, Denmark: Roskilde University.

Pérez, L.A.B. 2009. *La Industria del Etanol en México.* January. http://www.ejournal.unam. mx/ecu/ecunam16/ECU001600606.pdf (accessed October 1, 2009).

Peters, P. 2003. *Cutting Losses: Cuba Downsizes Its Sugar Industry.* Arlington, VA: Lexington Institute. December. http://www.lexingtoninstitute.org/cutting-losses-cuba-down-sizes-its-sugar-industry (accessed July 2009).

Population Division, Department of Economic and Social Affairs, United Nations Secretariat. 2008. *World Population Prospects: The 2008 Revision.* http://esa.un.org/unpp (accessed October 16, 2010).

Price, L., Worrell, E., Sinton, J., and Yun, J. 2001. *Industrial Energy Efficiency Policy in China.* Paper presented at the 2001 ACEEE Summer Study on Energy Efficiency in Industry in Tokyo.

PSA. 2009. *Citroen Reveals the C-Zero, a 100% Electric City Car.* Press Release. November 11. http://www.psa-peugeot-citroen.com/en/news/tmp_innovation_details_a. php?id=585 (accessed April 10, 2010).

Qi, J., Zheng, Y., and Zhao, Y. 2007. Environmental Regulation and Trade Pattern: A Case of China. *Ecological Economy* 3:234–242.

Queiroz, A.U.B., and Collares-Queiroz, F.P. 2009. Innovation and Industrial Trends in Bioplastics. *Journal of Macromolecular Science, Part C: Polymer Reviews* 49:65–78.

Rapoza, K. 2008. Brazil's Bet: Ethanol Surpasses Gasoline in Brazil. *Wall Street Journal.* September 19.

Ravinovitch, J. 1992. Curitiba: Towards Sustainable Urban Development. *Environment and Urbanization* 4(2):62–73.

Reddy, S., and Painuly, J.P. 2004. Diffusion of Renewable Energy Technologies—Barriers and Stakeholders' Perspectives. *Renewable Energy* 29:1431–1447.

Reinhardt, F.L. 2008. Environmental Product Differentiation: Implications for Corporate Strategy. In Russo, M.V. (Ed.), *Environmental Management: Readings and Cases.* Thousand Oaks, CA: Sage, pp. 205–227.

REN21. 2009. *Renewables Global Status Report: 2009 Update.* Paris: REN21 Secretariat. http://www.ren21.net/Portals/97/documents/GSR/RE_GSR_2009_Update.pdf

Renault. 2010. The Electric Vehicle: A Global Strategy. http://www.renault.com/en/capeco2/vehicule-electrique/pages/vehicule-electrique.aspx (accessed April 10, 2010).

Renewable Fuels Associates. 2010. Annual World Ethanol Production by Country. Renewable Fuels Association. http://www.ethanolrfa.org/pages/statistics.

Reuters. 2010. PSA, Mitsubishi partnership talks stall-newspaper. January 25. http://www. reuters.com/article/idUSLDE60O2JL20100125 (accessed April 10, 2010).

Rex, E., and Baumann, H. 2007. Beyond Ecolabels: What Green Marketing Can Learn from Conventional Marketing. *Journal of Cleaner Production* 15:567–576.

RFA (Renewable Fuels Association). 2008. *World Fuel Ethanol Production 2008*. http://www.ethanolrfa.org/industry/statistics/#E (accessed February 2009).

Rice-Oxley, M. 2007. Air Travel Latest Target in Climate Change Fight. *Christian Science Monitor*. August 17. http://www.csmonitor.com/2007/0817/p01s01-woeu.html (accessed November 5, 2009).

Risø National Laboratory. 2005. *Risø Energy Report 4: The Future Energy System. Distributed Production and Use—Summary*. October.

Risø National Laboratory. 2009. *Risø Energy Report 8: The Intelligent Energy System. Infrastructure for the Future—Summary*. September.

Rohter, L. 2006. With Big Boost from Sugar Cane, Brazil Is Satisfying Its Fuel Needs. *New York Times*. April 10.

Romero, S. 2009. To Protect Galápagos, Ecuador Limits a Two-legged Species. *New York Times*. October 5. Published online October 4. http://www.nytimes.com/2009/10/05/world/americas/05galapagos.html?_r=1&scp=1&sq=galapagos&st=cse (accessed November 5, 2009).

Roth, A., and Kåberger, T. 2002. Making Transport Systems Sustainable. *Journal of Cleaner Production* 10:361–371.

Rothkopf, G. 2007. *A Blueprint for Green Energy in the Americas*. Paper prepared for the Inter-American Development Bank. December.

Sánchez, J.T. 2008. *Cuba y el Etanol: Proyecciones para una Economía Privada*. Paper presented at the Seventeenth Annual Meeting of the Association for the Study of the Cuban Economy (ASCE).

Sarkar, A.U., and Karagoz, S. 1995. Sustainable Development of Hydroelectric Power. *Energy* 20:977–981.

Schäfer, A., and Jacoby, H.D. 2006. Vehicle Technology under CO_2 Constraint: A General Equilibrium Analysis. *Energy Policy* 34:975–985.

Schiffler, M. 2004. Perspectives and Challenges for Desalination in the 21st Century. *Desalination* 165:1–9.

Schipper, L. 1983. Residential Energy Use and Conservation in Denmark, 1965–1980. *Energy Policy*, 11(4):313–323.

Schmitz, H. 2004. *Local Enterprises in the Global Economy: Issues of Governance and Upgrading. Section 4: Environmental and Social Standards*. London: Institute of Development Studies, pp. 71–84.

Scott, D., Peeters, P., and Gössling, S. 2010. Can Tourism Deliver its "Aspirational" Greenhouse Gas Emission Reduction Targets? *Journal of Sustainable Tourism* 18(3):393–408.

Şen, Z. 2004. Solar Energy in Progress and Future Research Trends. *Progress in Energy and Combustion Science* 30:367–416.

SETRAVI. 2009. Mexico City Secretary of Transport and Roads. www.setravi.df.gob.mx.

Shannon, M.A., Bohn, P.W., Elimelech, M., Georgiadis, J.G., Mariñas, B.J., and Mayes, A.M. 2008. Science and Technology for Water Purification in the Coming Decades. *Nature* 452:301–310.

Shapouri H., and Salassi, M. 2006. *The Economic Feasibility of Ethanol Production from Sugar in the United States*. July, p. vi. Washington, DC: U.S. Department of Agriculture.

Small, K.A. 2009. Transportation: Urban Transportation Policy. In Inman, R.P. (Ed.), *Making Cities Work: Prospects and Policies for Urban America*. Princeton, NJ: Princeton University Press, pp. 63–93.

Smeets, E., Junginger, M., Faaij, A., Walter, A., and Dolzan, P. 2006. *Sustainability of Brazilian Bio-ethanol.* Report NWS-E-200-110. Utrecht and Campinas: Copernicus Institute at Universiteit Utrecht and Universidade Estadual de Campinas.

Smith, J.F. 2009. Harvard Honors Mexico City Bus System. *Boston Globe.* November 13.

Sperling, D., and Gordon, D. 2009. *Two Billion Cars: Driving toward Sustainability.* New York: Oxford University Press.

SRL. 2009. *Design for Recycling.* Georgia Institute of Technology, Systems Realization Laboratory. http://www.srl.gatech.edu/education/ME4171/DFR-Intro.ppt (accessed November 1, 2009).

Statistics Denmark. http://www.statistikbanken.dk/statbank5a/SelectVarVal/saveselections.asp (accessed April 2010).

Stebbins, C. 2007. Corn Ethanol Not the Culprit for Food Inflation. *Reuters.* December 10.

Steck, B., Strasdas, W., and Gustedt, E. 1999. Introduction. In *Sustainable Tourism as a Development Option: Practical Guide for Local Planners, Developers and Decision Makers*, R. Haep (Ed.), 1–3. Eschborn: Federal Ministry for Economic Co-operation and Development.

Steg, L. 2008. Promoting Household Energy Conservation. *Energy Policy* 36:4449–4453.

Stern, N. 2007. *The Economics of Climate Change: The Stern Review.* Cambridge: HM Treasury.

Stiglitz, J.E. 1997. *Stepping towards Balance: Addressing Global Climate Change.* October 6. Speech delivered at the Conference on Environmentally and Socially Sustainable Development, Washington, DC.

Subaru. 2009. *Fuji Heavy Industries (FHI) to Launch Subaru Plug-in Stella EV in Japan.* Press Release. June 4. http://www.fhi.co.jp/english/news/press/2009/09_06_04e.html (accessed April 10, 2010).

SwRI. 2009. *Efficiency Improvements Associated with Ethanol-Fueled Spark-Ignition Engines.* San Antonio, TX: Southwest Research Institute. April 23. http://www.swri.edu/4org/d03/engres/spkeng/sprkign/pbeffimp.htm (accessed July 14, 2009).

Sydorovych, O., and Wossink, A. 2008. The Meaning of Agricultural Sustainability: Evidence from a Conjoint Choice Survey. *Agricultural Systems* 98:10–20.

Syme, G. J., Shao, Q., Po, M., and Campbell, E. 2004. Predicting and Understanding Home Garden Water Use. *Landscape and Urban Planning* 68:121–128.

Tepelus, C.M. 2005. Aiming for Sustainability in the Tour Operating Business. *Journal of Cleaner Production* 13:99–107.

Tesla. 2008. Tesla Motors website. http://www.teslamotors.com (accessed April 10, 2010).

Tesla. 2010. *Telsa Motors and Toyota Motor Corporation Intend to Work Jointly on EV, TMC to Invest in Tesla.* Press Release. May 20. http://www.teslamotors.com/about/press/releases/tesla-motors-and-toyota-motor-corporation-intend-work-jointly-ev-development-tm

Toasa, J. 2009. *Colombia: A New Ethanol Producer on the Rise?* Washington, DC: U.S. Department of Agriculture.

Toyota. 2009a. *All New Third Generation Toyota Prius Raises the Bar for Hybrid Vehicles— Again.* Press Release. March 2. http://pressroom.toyota.com/pr/tms/toyota/all-new-third-generation-toyota-82970.aspx (accessed April 10, 2010).

Toyota 2009b. *Toyota, Lexus Hybrids Top 1 Million Sales in U.S.* Press Release. March 11. http://pressroom.toyota.com/pr/tms/toyota/toyota-and-lexus-hybrids-top-one-85047.aspx (accessed April 10, 2010).

UNESCO. 2009. *Water in a Changing World.* Paris: UNESCO.

United Nations General Assembly. 1987. *Report of the World Commission on Environment and Development: Our Common Future.* http://www.un-documents.net/wced-ocf.htm

UNWTO (World Tourism Organization). 2009. *UNWTO Tourism Highlights*, 2009 Edition. http://www.unwto.org/facts/eng/highlights.htm (accessed November 5, 2009).

Urbanchuk, J.M. 2009. *Contribution of the Ethanol Industry to the Economy of the United States.* Paper prepared for the Renewable Fuels Association. LEGG, LLC. February 23.

U.S. Department of Energy. 2006. *Trade and Integration.* Washington, DC: U.S. Department of Energy, Energy Information Administration. November. http://www.eia.doe.gov/emeu/cabs/chapter6a.html (accessed September 29, 2009).

U.S. Department of Energy. 2009. *International Energy Outlook 2009.* Washington, DC: U.S. Department of Energy, Energy Information Administration.

U.S. Department of Energy, Energy Information Administration. n.d. *Spot Prices.* Washington, DC: U.S. Department of Energy, Energy Information Administration. http://tonto.eia.doe.gov/dnav/pet/pet_pri_spt_s1_d.htm (accessed April 2010).

USDA Foreign Agricultural Service. 2003. *Brazil: Future Agricultural Expansion Potential Underrated.* January 21. http://www.fas.usda.gov/pecad2/highlights/2003/01/Ag_expansion/index.htm (accessed March 2009).

von Blottnitz, H., and Curran, M.A. 2007. A Review of Assessments Conducted on Bio-ethanol as a Transportation Fuel from a Net Energy, Greenhouse Gas, and Environmental Life Cycle Perspective. *Journal of Cleaner Production* 15:607–619.

von Medeazza, G.M., and Moreau, V. 2007. Modeling of Water-Energy Systems: The Case of Desalination. *Energy* 32:1024–1031.

Wandel, M., and Bugge, A. 1997. Environmental Concern in Consumer Evaluation of Food Quality. *Food Quality and Preference* 8:19–26.

Watkins, G., and Cruz, A. 2007. *Galapagos at Risk: A Socioeconomic Analysis of the Situation in the Archipelago.* Puerto Ayora, Galapagos, Ecuador: Charles Darwin Foundation.

WEF. 2009. *Green Investing: Towards a Clean Energy Infrastructure.* Davos, Switzerland: World Economic Forum.

Westbrook, M. 2001. *The Electric Car.* Warrendale, PA: Society of Automotive Engineers.

Worden, R.L., Savada, A.M., and Dolan, R.E. (Eds.) 1987. *China: A Country Study.* Washington, DC: GPO for the Library of Congress.

World Bank. 1998. *Questions and Answers on the World Bank and Climate Change.* Washington, DC: World Bank.

World Bank. 2000. *The World Development Report 1999/2000: Entering the 21st Century: The Changing Development Landscape, Volume 1.* New York: Oxford University Press.

World Bank. 2009. *Data Visualizer—World Development Indicators 2009.* http://devdata.worldbank.org/DataVisualizer (accessed April 2010).

World Bank. *Data Statistics.* http://web.worldbank.org/WEBSITE/EXTERNAL/DATASTATISTICS (accessed April 2010).

World Bank Group. 2009. *Sector Brief: Transportation in MENA.* Washington, DC: World Bank.

World Energy Council. 2007. *Denmark Energy Efficiency/CO_2 Indicators 2007.* http://www.worldenergy.org/documents/dnk.pdf (accessed April 2009).

World Energy Council. 2008. *Denmark Energy Efficiency/CO_2 Indicators 2008.* http://www.worldenergy.org/documents/dnk.pdf (accessed October 2010).

World Resources Institute. 2008. *National Alcohol Program (PROALCOOL).* http://projects.wri.org/sd-pams-database/brazil/national-alcohol-program-proalcool (accessed March 23, 2008).

Worrell, E., Laitner, J.A., Ruth, M., and Finman, H. 2003. Productivity Benefits of Industrial Energy Efficiency Measures. *Energy* 28:1081–1098.

WWF (World Wildlife Fund). 2009. http://www.worldwildlife.org (accessed November 5, 2009).

Xavier, M.R. 2007. *The Brazilian Sugarcane Ethanol Experience.* Working Paper for the Competitive Enterprise Institute. February 15.

Yaw, F., Jr. 2005. Cleaner Technologies for Sustainable Tourism: Caribbean Case Studies. *Journal of Cleaner Production* 13:117–134.

Yergin, D. 2009. *The Long Aftershock: Oil and Energy Security after the Price Collapse.* Testimony to the Joint Economic Committee of the U.S. Congress, Washington, DC, May 20.

Zhang, ZhongXiang. 2010. *Assessing China's Energy Conservation and Carbon Intensity: How Will the Future Differ from the Past?* May 24. Working Paper, East-West Center, Research Program, Honolulu, HI.

Appendix 1: The Green Products Universe

1. Clean and/or Renewable Energy Sources (Generation and Storage)
   Solar
   Wind
   Geothermal
   Biomass
   Hydro
   Ethanol

2. Energy Use
   Mass transit solutions
   Alternative automobile engines
   Efficient buildings: heating and air conditioning, insulation
   Industrial efficiency
   Lighting
   Household appliances

3. New Environmental and/or Recyclable Products and Materials
   Design to facilitate recycling: computers, cell phones, automobiles
   Bioplastics
   Packaging
   Recycling technologies and waste management

4. Water Management
   Green agriculture
   Gardening and landscaping
   Sewage systems
   Purification
   Desalination
   Chemical waste treatment

5. Lifestyles
   Food consumption habits
   Green leisure
   Green tourism
   Waste generation, recycling, and disposal

6. Environmental and Sustainability Policies
   Regulation
   Standards
   Certification
   Diffusion
   Subsidies

7. Adapting Business Practices
   Product design and development
   Operations (manufacturing or service)
   Accounting
   Marketing
   Organization

Appendix 2: Essential Readings on Green Products

CLEAN AND RENEWABLE ENERGY SOURCES

Renewable Energy in General

1. Reddy, S., and Painuly, J.P. Diffusion of renewable energy technologies—barriers and stakeholders' perspectives. *Renewable Energy* 29 (2004): 1431–1447. This survey summary develops systematic classification and ranking of barriers to the adoption of RETs (economic, technological, market, and institutional) and analyzes the results.
2. Lior, N. Energy resources and use: The present situation and possible paths to the future. *Energy* 33 (2008): 842–857. Summarizes the current status and future potential of world oil, gas, coal resources, nuclear, and renewable energy uses.

Solar

1. Şen, Z. Solar energy in progress and future research trends. *Progress in Energy and Combustion Science* 30 (2004): 367–416. This paper provides a comprehensive account of solar energy sources and conversion methods.
2. Kalogirou, S.A. Environmental benefits of domestic solar energy systems. *Energy Conversion and Management* 45 (2004): 3075–3092. This paper presents examples of pollution caused by conventional energy sources, compared with the environmental impacts of solar water heating and solar space heating systems.
3. Fthenakis, V., Mason, J.E., and Zweibel, K. The technical, geographical, and economic feasibility for solar energy to supply the energy needs of the U.S. *Energy Policy* 37 (2009): 387–399. This paper presents the solar energy potential in overall U.S. energy supply from both a technical and economic point of view. This can be considered an indicator of global solar energy potential.

Wind

1. Karki, R. Renewable energy credit driven wind power growth for system reliability. *Electric Power Systems Research* 77 (2007): 797–803. This paper presents useful information and an evaluation method of the wind energy application in electric power systems.

Geothermal

1. Fridleifsson, I.B. Present status and potential role of geothermal energy in the world. *Renewable Energy* (1996): 34–39. Special Issue World Renewable Energy Congress (WREC). This paper summarizes the status of the applications of geothermal energy and their world distribution. It concludes that geothermal energy has the largest electricity-related potential.
2. Gallup, D.L. Production engineering in geothermal technology: A review. *Geothermics* 38 (2009): 326–334. This paper reviews the production engineering technology in the use of geothermal as a competitive renewable energy resource.

Biomass

1. McKendry, P. Energy production from biomass. *Bioresource Technology* 83 (2002): 37–63. This review paper is divided into three parts: Overview of biomass, conversion technologies, and gasification technologies. It provides a comprehensive review of biomass energy use.
2. Field, C.B., Campbell, J.E., and Lobell, D.B. Biomass energy: The scale of the potential resource. *Trends in Ecology and Evolution* 23 (2008): 65–72. This paper gives a comprehensive review of the potential of biomass as a useable energy resource from several critical aspects.

Hydro

1. Sarkar, A.U., and Karagoz, S. Sustainable development of hydro-electric power. *Energy* 20 (1995): 977–981. This paper discusses the environmental problems created by increasing demand for electricity. From several examples, it concludes in favor of small hydroelectric plants.

2. Glasnovic, Z., and Margeta, J. The features of sustainable solar hydroelectric power plant. *Renewable Energy* 34 (2009): 1742–1751. This paper presents the main features of the new power plant technology combining a modified reversible hydroelectric (HE) power plant with photovoltaic (PV) power plant. This SHE represents a permanently sustainable energy source.

Ethanol

1. von Blottnitz, H., and Curran, M.A. A review of assessments conducted on bio-ethanol as a transportation fuel from a net energy, greenhouse gas, and environmental life cycle perspective. *Journal of Cleaner Production* 15 (2007): 607–619. It provides analysis of the key results of 47 assessments that compare bio-ethanol systems to conventional. Bio-ethanol does result in reductions in both resource use and global warming, but many other negative impacts are more often unfavorable than favorable.
2. Goldemberg, J., Coelho, S.T., and Guardabassi, P. The sustainability of ethanol production from sugarcane. *Energy Policy* 36 (2008): 2086–2097. This article examines the sustainability of ethanol production in Brazil covering environmental aspects (including air, water, land use, soil, and biodiversity) and social aspects (including jobs, income, and working conditions).

ENERGY USE

Mass Transit Solutions

1. Litman, T. Efficient vehicles versus efficient transportation. Comparing transportation energy conservation strategies. *Transport Policy* 12 (2005): 121–129. This paper introduces an evaluation framework to quantitatively analyze and compare four strategies to reduce transport energy consumption. The factors taken into account are comprehensive.

2. Pro, B.H., Hammerschlag, R., and Mazza, P. Energy and land use impacts of sustainable transportation scenarios. *Journal of Cleaner Production* 13 (2005): 1309–1319. In this study, energy efficiency of transportation is calculated for four hypothetical, renewable fuel cycles. The overall system efficiency is analyzed based on energy and land use value.

3. Roth A., and Kåberger T. Making transport systems sustainable. *Journal of Cleaner Production* 10 (2002): 361–371. This report used different companies to illustrate the management schemes to control CO_2 emissions in transport systems.

Alternative Automobile Engines

1. Åhman, M. Government policy and the development of electrical vehicles in Japan. *Energy Policy* 34 (2006): 433–443. Analyzes the impacts of government on the development of alternatives to conventional vehicles, focused on battery-powered electric vehicles (BPEVs) and concludes that government should have a positive role in technological innovation.

2. Schäfer, A., and Jacoby, H.D. Vehicle technology under CO_2 constraint: A general equilibrium analysis. *Energy Policy* 34 (2006): 975–985. This article presents a study of the market penetration of different transport technologies under policy constraints on CO_2 emissions.

Efficient Buildings: Heating and Air Conditioning, Insulation

1. John, G., Clements-Croome, D., and Jeronimidis, G. Sustainable building solutions: A review of lessons from the natural world. *Building and Environment* 40 (2005): 319–328. This paper develops the concept of sustainable buildings and explores the relationship to the environment from building materials and design aspects.

2. Xia, C., Zhu, Y., and Lin, B. Renewable energy utilization evaluation method in green buildings. *Renewable Energy* 33 (2008): 883–886. This paper presents a better way to assess renewable energy utilization, an important part of green buildings, using energy quality coefficient (EQC) and energy conversion coefficient (ECC).

Industrial Efficiency

1. Ayres, R.U., Turton, H., and Casten, T. Energy efficiency, sustainability and economic growth. *Energy* 32 (2007): 634–648. This article explores the relationship between energy efficiency of industrial systems and processes and economic growth.
2. Worrell, E., Laitner, J.A., Ruth, M., and Finman, H. Productivity benefits of industrial energy efficiency measures. *Energy* 28 (2003): 1081–1098. This paper reviews the relationship between energy improvement measures and productivity in industry through case studies and detailed analysis.

Lighting

1. Loe, D.L. Quantifying lighting energy efficiency: A discussion document. *Lighting Research & Technology* 35,4 (2003): 319–329. This paper aims to establish a quantification system to determine the energy efficiency of lighting in various situations, such as equipment, installation design, lighting use, and design approach.

Household Appliances

1. Dzioubinski, O., and Chipman, R. Trends in consumption and production: Household energy consumption. DESA Discussion Paper No. 6 (1999). This paper summarizes the current household energy consumption situation and contains valuable data and information.
2. Steg, L. Promoting household energy conservation. *Energy Policy* 36 (2008): 4449–4453. This paper presents the current state of factors influencing household energy use, and then examines the strategies to promote household energy savings.

NEW ENVIRONMENTAL AND RECYCLABLE PRODUCTS AND MATERIALS

Design to Facilitate Recycling

1. Lee, C.H., Chang, C.T., Fan, K.S., and Chang, T.C. An overview of recycling and treatment of scrap computers. *Journal of Hazardous Materials* B114 (2004): 93–100. This article reviews the recycling and treatment of used computers at a technical level, giving details of the important processes and methods adopted.
2. Williams, E., Kahhat, R., Allenby, B., Kavazanjian, E., Kim, J., and Xu, M. Environmental, social, and economic implications of global reuse and recycling of personal computers. *Environmental Science and Technology* 42 (2008): 6446–6454. This paper presents the big picture of reuse and recycling of used computers from environmental, economic, and social perspectives and concludes that more policies and technologies need to be developed.
3. Reedy, S. A wireless waste. *Telephony* (2008): 12. This is a short article about the current situation of cell phone recycling.
4. Ferrao, P., and Amaral, J. Design for recycling in the automobile industry: New approaches and new tools. *Journal of Engineering Design* 17 (2006): 447–462. This paper presents novel design for recycling (DFR) strategies for automobiles, which combine the technologies of dismantling and shredding, the two main industries of end-of-life vehicle processing.

Bioplastics

1. Queiroz, A.U.B., and Collares-Queiroz, F.P. Innovation and industrial trends in bioplastics. *Journal of Macromolecular Science, Part C: Polymer Reviews* 49 (2009): 65–78. This paper provides a brief summary of the scenario and past developments, recent data of different types of bioplastics, and an analysis of future trends.

Packaging

1. Kale, G., Kijchavengkul, T., Auras, R., Rubino, M., Selke, S., and Singh, S.P. Compostability of bioplastic packaging materials: An overview. *Macromolecular Bioscience* 7 (2007): 255–277. This paper first summarizes the current situation of packaging materials and related waste problems and then focuses on bioplastic packaging.

Recycling Technologies and Waste Management

1. Giusti, L. A review of waste management practices and their impact on human health. *Waste Management* 29 (2009): 2227–2239. This paper presents a comprehensive review of waste generation, management practices, and related health issues.
2. Nowosielski, R., and Zajdel, A. Recycling's technology. *Journal of Achievements in Materials and Manufacturing Engineering* 21 (2007): 85–88. This short article discusses the importance of recycling as a strategy to minimize waste, especially waste electrical and electronic equipment (WEEE).

WATER MANAGEMENT

Green Agriculture

1. Sydorovych, O., and Wossink, A. The meaning of agricultural sustainability: Evidence from a conjoint choice survey. *Agricultural Systems* 98 (2008): 10–20. This paper analyzes the results from a survey to investigate factors in determining agricultural sustainability.
2. Khan, S., Tariq, R., Yuanlai, C., and Blackwell, J. Can irrigation be sustainable? *Agricultural Water Management* 80 (2006): 87–99. This paper explores the irrigation problems, soil salinity, and low water-use efficiency issues resulting in the loss of agricultural lands. It also presents approaches to solve the environmental problems and maintain agricultural sustainability.

Gardening and Landscaping

1. Syme, G.J., Shao, Q., Po, M., and Campbell, E. Predicting and understanding home garden water use. *Landscape and Urban Planning* 68 (2004): 121–128. Introduces the importance of home gardens in current lifestyle and analyzes the water use in accordance with household attitudes and socio-demographic variables.
2. *Weather Reach—Landscape Water Management.* North Logan, UT: Irrisoft Inc. This training manual explains important issues related with landscape water management and how to integrate the factors in systems.

Sewage Systems

1. Burkhard, R., Deletic, A., and Craig, A. Techniques for water and wastewater management: A review of techniques and their integration in planning. *Urban Water* 2 (2000): 197–221. This paper presents a review of techniques in urban wastewater management, specifically in rainwater management, domestic wastewater management, and water re-use as well as their future potential.
2. Gogate, P.R., and Pandit, A.B. A review of imperative technologies for wastewater treatment I: Oxidation technologies at ambient conditions. *Advances in Environmental Research* 8 (2004): 501–551. Gogate, P. R. and Pandit, A.B. A review of imperative technologies for wastewater treatment II: Hybrid methods. *Advances in Environmental Research* 8 (2004): 553–597. This two-article series on technologies for wastewater treatment focuses on two kinds of methods and provides a comprehensive review of this field.

Purification

1. Shannon, M.A., Bohn, P.W., Elimelech, M., Georgiadis, J.G., Mariñas, B.J., and Mayes, A.M. Science and technology for water purification in the coming decades. *Nature* 452 (2008): 301–310. This article gives a review of the science and technologies being developed for water purification.

Desalination

1. von Medeazza, G.M., and Moreau, V. Modeling of water-energy systems. The case of desalination. *Energy* 32 (2007): 1024–1031. This article assesses the sustainability of water desalination technologies through a model and concludes with potential environmental and social impacts of water desalination.
2. Schiffler, M. Perspectives and challenges for desalination in the 21st century. *Desalination* 165 (2004): 1–9. This article provides a global review of water desalination, including the technologies, framework, and challenges.
3. Khawaji, A.D., Kutubkhanah, I.K., and Wie, J.-M. Advances in seawater desalination technologies. *Desalination* 221 (2008): 47–69. This is a review of recent seawater desalination technologies with special focus on the multi-stage flash distillation and reverse osmosis processes.

LIFESTYLE

Food Consumption Habits

1. Halkier, B. Risk and food: Environmental concerns and consumer practices. *International Journal of Food Science and Technology* 36 (2001): 801–812. This paper presents a study about environmental risk-handling in food consumption practices in relation to the tension between individual consumer desire and control.
2. Wandel, M., and Bugge, A. Environmental concern in consumer evaluation of food quality. *Food Quality and Preference* 8 (1997): 19–26. This study discusses the environmental factors in consumers' evaluation of food quality and how they affect consumer's behaviors.

Leisure and Tourism

1. Tepelus, C.M. Aiming for sustainability in the tour operating business. *Journal of Cleaner Production* 13 (2005): 99–107. This paper investigates what the tour operating industry considers to be "good practice" and explores the sufficiency of these practices for sustainable tourism.

2. Choi, H.C., and Sirakaya, E. Sustainability indicators for managing community tourism. *Tourism Management* 27 (2006): 1274–1289. This study developed a set of 125 indicators for sustainable community tourism measurement in political, social, ecological, economic, technological, and cultural dimensions, which can serve as standards for tourism sustainability evaluation.

Waste Generation, Recycling, and Disposal

1. Lombrano, A. Cost efficiency in the management of solid urban waste. *Resources, Conservation and Recycling* 53 (2009): 601–611. This paper provides very detailed data to analyze the cost efficiency of waste management based on several factors such as technology, population, and waste collection.
2. Giusti, L. A review of waste management practices and their impact on human health. *Waste Management* 29 (2009): 2227–2239. This work reviews the most recent information on waste generation and waste disposal options and the potential impacts of these practices on health.

ENVIRONMENTAL AND SUSTAINABILITY POLICIES

Regulation

1. Shin, D., Curtis, M., Huisingh, D., and Zwetsloot, G.I. Development of a sustainability model for promoting cleaner production: A knowledge integration approach. *Journal of Cleaner Production* 16 (2008): 1823–1837. This paper presents an integrated knowledge approach and tests three functional modes of knowledge (contextual, technological, reconciliatory) and points out that fundamental disciplines of human knowledge provide the sufficient solutions to complex problems of sustainability.
2. Fischer, C., and Newell, R. Environmental and technology policies for climate change and renewable energy. Resources for the Future Discussion Paper 04-05 (2004). This paper assesses different policy options for reducing greenhouse gas emissions and promoting the development and diffusion of renewable energy technologies.

3. Heyes, A. Implementing environmental regulation: Enforcement and compliance. *Journal of Regulatory Economics* 17 (2000): 107–129. This research article addresses the enforcement and compliance issues in practicing environmental regulations by governments and organizations worldwide.

Standards

1. Schmitz, H. *Local Enterprises in the Global Economy: Issues of Governance and Upgrading.* Cheltenham: Edward Elgar (2004). Section 4: Environmental and Social Standards, pp. 71–84. This section of the book talks about the situation and development of environmental standards and how it affects the global economy.
2. Jabareen, Y. A new conceptual framework for sustainable development. *Environment Development and Sustainability* 10 (2008): 179–192. This literature review uses detailed conceptual analysis and arrives at seven concepts that assemble the framework for sustainable development.

Diffusion

1. Montalvo, C. General wisdom concerning the factors affecting the adoption of cleaner technologies: A survey. 1990–2007. *Journal of Cleaner Production* 16S1 (2008): S7–S13. This article presents a selective survey of papers to clarify the challenges facing diffusion modelers and policymakers, the assessment of the levels of diffusion achieved, and factors affecting the adoption of cleaner technologies.
2. Montalvo, C., and Kemp, R. Cleaner technology diffusion: Case studies, modeling and policy. *Journal of Cleaner Production* 16S1 (2008): S1–S6. This short paper is the introduction of a special issue, and it summarizes the state of the current cleaner technology diffusion market, research, and future opportunities.

Subsidies

1. Aalbers, R., van der Heijden, E., Potters, J., van Soest, D., and Vollebergh, H. Technology adoption subsidies: An experiment with managers. *Energy Economics* 31 (2009): 431–442. This experiment evaluates the impact of technology subsidies on the firm's behavior, especially investment decisions, and concludes that subsidies result in an increase in new technology adoption.
2. Olsson, L.E., and Gärling, T. Staying competitive while subsidized: A governmental policy to reduce production of environmentally harmful products. *Environment and Planning C: Government and Policy* 26 (2008): 667–677. This is a test on the governmental policy system proposed to reduce environmentally harmful products and concludes that it will not eliminate competitiveness.

ADAPTING BUSINESS PRACTICES

Product Design and Development

1. Reinhardt, F.L. Environmental Product Differentiation—Implications for Corporate Strategy. California Management Review 40 (1998): 43–73. This specifically talks about environmental product development and business practices.
2. Baumann, H., Boons, F., and Bragd, A. Mapping the green product development field: Engineering, policy and business perspectives. *Journal of Cleaner Production* 10 (2002): 409–425. This literature review of environmental product development (EPD) analyzes in detail product design and the environment from business, engineering, and policy perspectives.

Accounting

1. Jasch, C. Environmental management accounting (EMA) as the next step in the evolution of management accounting. *Journal of Cleaner Production* 14 (2006): 1190–1193. Provides information and analysis about the background, importance, basic methods, and current development of environmental management accounting (EMA).

2. Joshi, S., Krishman, R., and Lave, L. Estimating the hidden costs of environmental regulation. *The Accounting Review* 76 (2006): 171–198. Examines the cost of environmental regulation and the relation between visible costs and hidden costs and concludes that hidden costs can be pervasive.

Marketing

1. Rex, E., and Baumann, H. Beyond ecolabels: What green marketing can learn from conventional marketing. *Journal of Cleaner Production* 15 (2007): 567–576. Provides an introduction to basic ideas in green marketing and analyzes the current problems and inefficiency. The paper then introduces conventional marketing theories and explains how to combine these two to further improve green marketing.
2. Bonini, S.M., and Oppenheim, J.M. Helping "green" products grow. *The McKinsey Quarterly* (2008): 1–8. This paper presents how businesses can help change consumers' behavior by breaking down barriers to realize the green product's true potential.

Organization

1. Albino, V., Balice, A., and Dangelico, R.M. Environmental strategies and green product development: An overview on sustainability-driven companies. *Business Strategy and the Environment* 18 (2009): 83–96. This paper investigates the impact of environmental strategies adopted by sustainability-driven companies on the development of green products, and its economic or geographical differences.
2. Jones, M.T. Social responsibility and the "green" business firm. *Industrial and Environmental Crisis Quarterly* 9 (1996): 327–345. This paper relates social responsibilities and environmentalism and discusses their impacts on business firms' behavior.

Index

A

Accounting practices, 11
Acid rain, 89
Air pollutants, known cause of, 88
Air travel, 136
American Clean Energy and Security
    Act, 27
Automobile operation costs, 63

B

Battery
    -charging stations, 59
    disposal, 13
    gasoline tank, 58
    hydrogen tank, 58
    lithium-ion, 71
    nickel-metal hydride, 52
    -powered electric vehicles (BPEVs), 5,
        192, *see also* Japan, revival of
        battery-powered electric vehicles
        in
    -swapping stations, 61
    technology, advances, 56
BCG, *see* Boston Consulting Group
Bioenergetics Promotion and
    Development Law (Mexico), 130
Biomass, 187
    absolute demand, 22
    CHP electricity generated with, 43
    controversy, 4
    conversion product, 4
    essential readings, 190
    grid parity and, 32
    power generation capacity, 4
    production, definition of, 4
Bioplastics, definition of, 7
Boston Consulting Group (BCG), 61
BPEVs, *see* Battery-powered electric
    vehicles
Brazilian ethanol, 119–123
BRT, *see* Bus rapid transit

Bush, George W., 53
Bus rapid transit (BRT), 90
    attributes, 90
    characteristics, 90
    competition, 92
    conditions for success, 168
    cost, 92
    Mexico City, 82

C

Carbon capture and sequestration (CCS),
    114
Carbon dioxide emissions, 77
Carbon emissions intensity, 37
Carbon monoxide, 88
Cash-for-clunkers U.S. government-
    sponsored incentive, 61
Castro, Fidel, 132
CCS, *see* Carbon capture and
    sequestration
CHP, *see* Combined heat and power
Clean-tech powertrain technologies, 58
Climate change, *see also* Global warming
    danger of, 165
    GDP and, 25
    increased evidence of, 39
    policies, 27
        competitiveness, 27
        major, 27
Colombia (new ethanol producer),
    124–125
Combined heat and power (CHP), 42
    cooperative groups, 44
    electricity feed-in tariffs, 43
    gas, 40
    plants, 42
Competition
    Brazil, 119, 122
    BRT, 92
    Cairo, 104
    cities in developing countries, 95
    climate change policies, 27

Cuba, 132
Denmark energy plans, 40
Ecuador Ministry of Tourism, 139
energy efficiency, 30
EV TCO, 62
geothermal, 190
green powertrain technologies, 58
Japanese EV producers, 53
Mexican cities, 97
multilateral collaboration and, 46
Nissan, 61
nuclear energy, 32
subsidies, 200
sugar-based ethanol, 15
Compressed natural gas vehicles, 5
Corn ethanol, 14, 132
Cuba (potential sugar-based ethanol
    entrant), 130–133

D

Darwin, Charles, 142
Darwin Foundation, 153
Denmark, transition from oil dependency
    to sustainability in, 35–47
  biomass, CHP electricity generated
    with, 43
  carbon emissions intensity, 37
  competition, 40, 46
  Electricity Supply Act, 40
  energy intensity, definition of, 37
  "Energy 2000" plan, 39
  energy saving campaigns, 39
  energy scarcity, 35
  foreign oil, dependence on, 39
  fuel substitution, 39
  fuel supply, diversification, 36
  "green tax" package, 42
  Heat Supply Act, 39, 40, 42
  incentive system, 47
  key elements of Danish energy system,
    38–46
    development of network of
      combined heat and power
      plants, 42–43
    development of wind sector, 43–46
    energy efficiency and conservation,
      41–42
    energy policy continuity, 38–41

labeling programs, 41
large-scale electric plants, 42
legislation, 40
Natural Gas Supply Act, 40
overview, 36–39
  decoupling of economic growth
    from energy consumption, 37
  development of new clean
    technology industry, 38
  reduction of energy needs,
    foreign dependency, and CO_2
    emissions, 37
political agenda, 44
subsidies, 45
sustainability policies, 36
wind industry, 46
Desalination, 9
Design for Recycling (DFR) technologies,
    6
DFR technologies, *see* Design for
    Recycling technologies

E

Economic sustainability, 166
Eco-tourism, 137
Electric automobile engines, case study,
    13–14
Electric vehicles (EVs), 54–58, *see also*
    Japan, revival of battery-
    powered electric vehicles in
  cost to operate, 64
  global warming and, 55
  main reasons for rising demand for
    electric vehicles, 55–58
  maintenance costs, 64
  motors, 54
  negatives, 57
  product characteristics, 54–55
  superiority, 77
  total ownership cost advantages, 62–64
    vehicle maintenance costs, 64
    vehicle operating costs, 62–64
Electricity
  demand, 31
  infrastructure, 30
  storage, price of, 31
Electricity Supply Act (Denmark), 40
El Mirador, 155

ELV, *see* End-of-life vehicle
EMA, *see* Environmental management
    accounting
End-of-life vehicle (ELV), 6
Energy, *see also* Sustainable energy
        systems, need for
    biomass, 4
    demand, 19, 20
    efficiency
        competition, 30
        definition of, 4
    geothermal, 3
    green, most controversial source, 4
    grid, upgrading of, 31
    intensity, definition of, 37
    nuclear, 32
    provision, 166
    saving campaigns, 39
    scarcity, 35
    solar, 3
    supply, current levels of, 20
    systems, 13
    use, essential readings, 191–193
Environmental management accounting
        (EMA), 12
Essential readings, 189–201
    business practices, adapting, 200–202
        accounting, 200–201
        marketing, 201
        organization, 201
        product design and development,
            200
    clean and renewable energy sources,
            189–191
        biomass, 190
        ethanol, 191
        geothermal, 190
        hydro, 190–191
        renewable energy in general, 189
        solar, 189
        wind, 190
    energy use, 191–193
        alternative automobile engines, 192
        efficient buildings (heating and air
            conditioning, insulation), 192
        household appliances, 193
        industrial efficiency, 193
        lighting, 193
        mass transit solutions, 191–192

    environmental and sustainability
        policies, 198–200
        diffusion, 199
        regulation, 198–199
        standards, 199
        subsidies, 200
    lifestyle, 197–198
        food consumption habits, 197
        leisure and tourism, 197–198
        waste generation, recycling, and
            disposal, 198
    new environmental and recyclable
        products and materials, 194–195
        bioplastics, 194
        design to facilitate recycling, 194
        packaging, 195
        recycling technologies and waste
            management, 195
    water management, 195–197
        desalination, 197
        gardening and landscaping, 196
        green agriculture, 195
        purification, 196
        sewage systems, 196
Ethanol, 4, *see also* Sugarcane ethanol,
        promise of as cleaner
        combustion engine fuel
    advantages, 134
    agave, 129
    cellulosic, 113
    corn, 14, 132
    Cuba, 130
    essential readings, 191
    flex-fuel vehicle, 58
    fuels, case study, 14–15
    government policies (Brazil), 170
    import tariff, 127
    Mexican, 128, 129
    net effect on global warming, 4
    powertrain technologies and, 58
    production method, 4
    readings, 191
    world capacity of ethanol production,
        4
EVs, *see* Electric vehicles

F

Fertilizers, 130

FHI, *see* Fuji Heavy Industries
Flex-fuel vehicle, 58
Food consumption patterns, 9
Ford Model T, 112
Ford Motor Company, 52
Fossil fuels, 55
   -based transportation, 76
   biomass energy and, 4
Fuel cell electric vehicles, 5
Fuel-cell vehicle, 58
Fuel substitution, 39
Fuji Heavy Industries (FHI), 50

G

Galapagos National Park, *see* Tourism
Gasoline
   consumption, 119
   ethanol in, 4, 14, 134
   flex-fuel vehicle, 58
   –internal combustion engine-powered
      automobile, 51
   prices, 52, 62, 117
   production, 52
   tank battery, 58
GDP, *see* Gross domestic product
General Motors, 52
Geothermal energy, 3, 190
Global warming, *see also* Climate change
   beef production and, 9
   biomass energy and, 4
   clean and renewable energy sources
      and, 2
   effect of ethanol fuel on, 4
   electric vehicles and, 55
   ethanol and, 191
   mitigation, 55
   need for swift action against, 134
Government
   biofuel mandates (Mexico), 134
   climate change debates, 25
   intervention, 10
   -sponsored incentive, 61
   subsidies, 3, 11
Green agriculture, 8
Green buildings, 5
Green business firm, definition of, 12
Green energy source, most controversial, 4
Green gardening and landscaping, 8

Greenhouse gas emissions, 56
Green household appliances, 6
Green lighting, 5
Green manufacturing processes, 5
Green marketing, 12
Green packaging, 7
Green plastics, 7
Green product innovation and adoption,
      165–171
   alternative renewable energy, 168
   BRT system, 168
   durable consumer products, 166
   economic sustainability, 166
   energy provision, 166
   hidden costs, 166
   incentive alignment, 168–169
   large-scale services, 166
   leading globally and acting locally, 170
   leisure infrastructure, 166
   looking toward the future, 170–171
   market mechanisms, 166
   "people" component, 166
   "planet" component, 166
   population growth, 165
   public-private partnerships, 169
   renewable fuel, 166
   social sustainability, 166
   sustainable development, definition of,
      166
   systemic approach, 167–168
   top-down mandate, 170
   triple bottom line, 167, 169
Green products, essential readings,
      189–201
   business practices, adapting, 200–202
      accounting, 200–201
      marketing, 201
      organization, 201
      product design and development,
         200
   clean and renewable energy sources,
      189–191
      biomass, 190
      ethanol, 191
      geothermal, 190
      hydro, 190–191
      renewable energy in general, 189
      solar, 189
      wind, 190

energy use, 191–193
    alternative automobile engines, 192
    efficient buildings (heating and air
        conditioning, insulation), 192
    household appliances, 193
    industrial efficiency, 193
    lighting, 193
    mass transit solutions, 191–192
environmental and sustainability
        policies, 198–200
    diffusion, 199
    regulation, 198–199
    standards, 199
    subsidies, 200
lifestyle, 197–198
    food consumption habits, 197
    leisure and tourism, 197–198
    waste generation, recycling, and
        disposal, 198
new environmental and recyclable
        products and materials, 194–195
    bioplastics, 194
    design to facilitate recycling, 194
    packaging, 195
    recycling technologies and waste
        management, 195
water management, 195–197
    desalination, 197
    gardening and landscaping, 196
    green agriculture, 195
    purification, 196
    sewage systems, 196
Green products, universe of, 1–15, 187–188
    accounting practices, 11
    adapting business practices to green
        production, 11–12
    battery disposal, 13
    beef production, 9
    biomass
        conversion product, 4
        power generation capacity, 4
        production, 4
    bioplastics, definition of, 7
    case studies of green production and
        consumption, 12–15
        electric engines for automobiles,
            13–14
        energy systems, 13
        ethanol fuels, 14–15

green tourism, 15
    urban mass transit, 14
chemical feedstock, 4
clean and renewable energy sources, 2
commercialized innovations, 1
competition, 15
compressed natural gas vehicles, 5
congestion reduction, 5
desalination, 9
design for the environment, 2
Design for Recycling technologies, 6
diffusion of green practices, 10
dominant disposal method, 7
eco-design, 2
economic cycles, 3
end-of-life vehicle, 6
energy efficiency, definition of, 4
environmental management
        accounting, 12
ethanol, 4
    net effect on global warming, 4
    world capacity of ethanol
        production, 4
food consumption patterns, 9
fossil fuels, 4
geothermal energy, 3
government subsidies, 11
green agriculture, 8
green buildings, 5
green business firm, definition of, 12
green energy sources, 2–4
green energy use, 4–6
green fuel, 15
green gardening and landscaping, 8
green household appliances, 6
green lighting, 5
green manufacturing processes, 5
green marketing, 12
green packaging, 7
green plastics, 7
green policies and regulation, 10–11
green product, common definition of,
    2
green recycling practices, 6
green regulations, 10
green standards, 10
green tourism, 9
green vehicle, 5
green water management, 7–9

green water treatment, 8
gross domestic product, 7
heavy metal elements, 6
hydroelectric power, 3
internal combustion engine
    alternatives, 5
landfill, 7
learning curve effects, 2
lifestyles and green consumption, 9–10
main focus of research, 5
marketing strategies, 12
mass transit, 5
medium-term mismatch, 13
multi-effect distillation, 9
multi-stage flash distillation, 9
municipal solid waste, 7
new environmental and recyclable
    products and materials, 6–7
non-biodegradable materials, 6
non-governmental organizations, 10
organizational changes, 12
packaging waste, 7
reserve osmosis, 9
solar energy, 3
waste associated with electrical and
    electronic equipment, 6
water purification, 8
wind power, 3
zero-emission vehicles, 13
Green recycling practices, 6
Green tourism, 9, 15
Green vehicle, 5
Green water treatment, 8
Gross domestic product (GDP), 7, 19, 135
    Cairo, 103
    climate change, 25
    Danish economy, 37
    energy demand, 19, 20
    green water management, 7
    oil addiction, 22, 23

H

Heat Supply Act (Denmark), 39, 40, 42
Heavy metal elements, 6
HV, see Hybrid vehicles
Hybrid electric vehicles, 5
Hybrid vehicles (HVs), 50
    cost to operate, 64

distances driven, 57
long-distance travelers, 76
major mass producers of, 50
motors, 54
pricing, 50
skepticism toward, 72
Hydroelectric power, 3, 190–191
Hydrogen tank battery, 58

I

IEA, see International Energy Agency
Intergovernmental Panel on Climate
    Change (IPCC), 25
Internal combustion engine alternatives, 5
International Energy Agency (IEA), 18, 25
IPCC, see Intergovernmental Panel on
    Climate Change

J

Japan, revival of battery-powered electric
    vehicles in, 49–79
    automobile operation costs, 63
    battery-charging stations, 59
    battery recharging, 60
    battery-swapping stations, 61
    carbon dioxide emissions, 77
    case studies, 65–74
        Mitsubishi i-MiEV, 67–72
        Nissan Leaf, 72–74
        Subaru Stella Plug-in, 65–66
    cash-for-clunkers U.S. government-
        sponsored incentive, 61
    challenges in mass-marketing electric
        vehicles, 59–62
    clean-tech powertrain technologies, 58
    competition, 53, 58, 61
    electric vehicle total ownership cost
        advantages, 62–64
        vehicle maintenance costs, 64
        vehicle operating costs, 62–64
    ethanol, powertrain technologies and,
        58
    fast-charging station, 54
    flex-fuel vehicle, 58
    Ford Motor Company, 52
    fossil fuels, 55, 76
    fuel-cell vehicle, 58

General Motors, 52
greenhouse gas emissions, 56
history of electric cars, 51–53
hybrid vehicles, 50
internal combustion engine vehicles, 59, 60
Japanese Road Transportation Vehicle Law, 50
Lithium Energy Japan, 69
lithium-ion battery, 71
Mitsubishi, 67, 69, 70
Nissan Auto Loan, 74
Nissan Leaf, 53
oil reserves, discovery of in Texas, 52
overview of electric vehicles, 54–58
    main reasons for rising demand for electric vehicles, 55–58
    product characteristics, 54–55
plug-in hybrid, 58
promotional initiatives, 62
specifications comparison table, 78
stopgap solution, electric cars as, 52
Subaru, 53, 65, 66
Tesla Motors, 53
Toyota Motor Corporation, 49, 53
vehicle-equipped inverter, 60
Japanese Road Transportation Vehicle Law, 50

K

Kyoto Protocol, 108

L

Landfill sites, 6, 7
Legislation
    American Clean Energy and Security Act, 27
    Electricity Supply Act (Denmark), 40
    Heat Supply Act (Denmark), 39, 40, 42
    Natural Gas Supply Act (Denmark), 40
LEJ, *see* Lithium Energy Japan
Lithium Energy Japan (LEJ), 69
Lithium-ion battery, 71

M

Manufacturer's suggested retail price (MSRP), 61
Mass transit, 5, *see also* Urban mass transport, sustainable
MED, *see* Multi-effect distillation
Mexico (ethanol and income equality), 128–130
Mitsubishi i-MiEV, 67–72
MSF, *see* Multi-stage flash distillation
MSRP, *see* Manufacturer's suggested retail price
MSW, *see* Municipal solid waste
Multi-effect distillation (MED), 9
Multi-stage flash distillation (MSF), 9
Municipal solid waste (MSW), 7

N

NAFTA partner, 118
National Program for Tourism Training, 139
Natural Gas Supply Act (Denmark), 40
NGO, *see* Non-governmental organization
Nickel-metal hydride (NiMH) battery, 52
NiMH battery, *see* Nickel-metal hydride battery
Nissan
    Auto Loan, 74
    competition, 61
    Leaf, 53, 72–74
Nitrogen dioxide, 89
Non-biodegradable materials, 6
Non-governmental organization (NGO), 10, 96
Nuclear energy, 32

O

Obama, Barack, 55, 63, 108, 128
OECD countries, 24, 83
OEMs, *see* Original equipment manufacturers
Oil
    addiction, 22
    dependency, *see* Denmark, transition from oil dependency to sustainability

exports, 23
futures, 117
-importing countries, policies of, 22
prices, 23
reserves, discovery of in Texas, 52
Optibus, 97
Organizational changes, 12
Original equipment manufacturers
(OEMs), 60, 62

P

Packaging waste, 7
PAHs, *see* Poly-aromatic hydrocarbons
Pesticides, 130
Petroleo Brasileiro, 116
Plug-in hybrid, 58
Poly-aromatic hydrocarbons (PAHs), 89
PPPs, *see* Public-private partnerships
Public-private partnerships (PPPs), 82

R

Renewable energy, essential readings, 189
Reserve osmosis (RO), 9
RO, *see* Reserve osmosis

S

Social sustainability, 166
Solar energy, 3, 189
Subaru Stella Plug-in, 65–66
Sugarcane ethanol, 14, 133, 169
Sugarcane ethanol, promise of as cleaner
combustion engine fuel, 107–134
airflow sensor feedback, 112
annual fuel ethanol production by
country, 116
Bioenergetics Promotion and
Development Law, 130
carbon capture and sequestration, 114
cellulosic ethanol, 113
corporate giants, 116
criticism, 113
"dead-zone" algae bloom, 129
economic and public policy issues,
117–119
environmental benefits, 123
ethanol

agave, 129
import tariff, 127
income equality (Mexico), 128–130
-only fuel, 111
feedstock candidates, 129
fertilizers, 130
flexible-fuel engine, 112
Ford Model T, 112
general considerations on ethanol,
110–114
important ethanol fuel
characteristics, 111–114
technology of ethanol fuel, 110–111
global reaction, 108
intervention, 122
Kyoto Protocol, 108
major producers and key players,
115–117
molasses treatment, 111
NAFTA partner, 118
National Alcohol Program, 120
new ethanol producer (Colombia),
124–125
nitrogen fertilizer, 125
oil futures, 117
pesticides, 130
Petroleo Brasileiro, 116
political barriers and market failure
(United States), 125–128
potential sugar-based ethanol entrant
(Cuba), 130–133
pro-alcohol biofuel promotion
program, 116
production steps, 110
role model (Brazilian ethanol),
119–123
rural poor, 130
subsidies, 122
sucrose yield, 115
sugarcane ethanol characteristics, 108
sugarcane production cost
classification, 117
tequila, 129
top sugarcane producing countries,
115
tortilla crisis, 129
transition to ethanol, 109
transitioning cost, 133
U.S. consumer price index, 126

value-added tax, 124
water-table contamination, 126
World Economic Forum, 107
Sulfur dioxide, 88
Sustainability, *see* Denmark, transition
    from oil dependency to
    sustainability
Sustainable development, definition of,
    166
Sustainable energy systems, need for,
    17–33
    assessment of attitudes and behaviors,
      25–29
      climate change, 25–26
      consumers, 28
      corporations, 28–29
      governments' sustainability
        policies, 26–27
    biomass, absolute demand, 22
    "business as usual" scenario, 25
    coal, 20
    electricity demand, 31
    energy efficiency, competition, 30
    GDP increase, 19
    grid parity, 32
    Intergovernmental Panel on Climate
      Change, 25
    International Energy Agency, 18, 25
    major climate change policies, 27
    nuclear energy, 32
    OECD countries, 24
    oil exports, 23
    oil-importing countries, policies of, 22
    oil prices, 23
    onshore wind power, 32
    population pyramid of developing
      countries, 19
    proposed transition guidelines, 29–32
      clean energy sources, 31–32
      electricity infrastructure, 30–31
      energy efficiency, 30
    rationale for sustainable change, 18–24
      energy demand, 19–20
      energy security, 24
      energy supply may not match
        future demand growth, 20–22
      future energy cost, 23–24
      oil addiction, 22
    renewable energy, 20

    supply-side measures, 30
    sustainable products, 28
Sustainable products, 28
Sustainable tourism, 137, *see* also Tourism

T

TCO, *see* Total cost of ownership
Tesla Motors, 53
Tortilla crisis, 129
Total cost of ownership (TCO), 62
Total suspended particles (TSPs), 87, 88
Tourism (sustainable), 135–164
    accreditations, 139
    air travel, 136
    archipelago, 138
    conservation groups, 152
    country of four worlds, 138
    cracked lava formations, 146
    Darwin Foundation, 153
    eco-tourism, 137
    Ecuador and the Galapagos islands,
      138–149
      characteristics of the Galapagos,
        140–142
      human involvement with the
        Galapagos, 142–149
    Ecuadorian economy, 139
    El Mirador, 155
    endemic land iguanas, 147
    ethanol, advantages, 134
    Galapagos Islands, 138, 140
      blue-footed boobies, 147
      flightless cormorant, 146
      giant tortoises, 147
      isolation, 141
      lava cactus, 142
      ocean-swimming marine iguanas,
        143
      penguin, 143
      red-footed boobies, 145
      satellite view, 141
      tourist categories, 151
      waved albatross, 148
    global warming, need for swift action
      against, 134
    government agencies, 153
    government biofuel mandates, 134
    incentive mechanism, 159

island-nations, 135
main issues as expressed by
        stakeholders, 155–162
    awareness of current situation, 156
    better education and stronger law
        enforcement, 162
    Ecuadorian (domestic) tourist
        vision misalignment, 159–160
    negative cycle (tourism driving
        migration, driving expansion,
        driving destruction), 161–162
    profitability and ethics, 158–159
    tensions between tour operators
        and retailers, 160–161
    tourists' central concern (cost),
        156–158
migration/population control, 155
National Program for Tourism
        Training, 139
once-active volcano, 147
PEMEX, 134
rock formations, 144
Sally lightfoot crabs, 146
salt-water lagoons and tide ponds, 145
sandy beaches and seals, 144
skyrocketed tourism, 145
"Special Law for Galapagos," 153
sustainability, 139
sustainable tourism, 137
tourism and tourists in the Galapagos,
        150–155
    current concerns, 153–155
    overview, 150–152
    stakeholders, 152–153
    tourist companies, 152
Toyota Motor Corporation, 49, 53
TSPs, *see* Total suspended particles

U

United Nations Bruntland Commission of
        1987, 166
United States (political barriers and
        market failure), 125–128
Urban mass transport, sustainable, 81–106
    acid rain, 89
    air pollutants, known cause of, 88
    below-cost pricing, 94
    carbon monoxide, 88

case study, 14, 95–102
    design, 97–98
    new business model, 98–99
    operational and economic
        efficiency, 100–101
    social and environmental impact,
        101–102
    solution, 96–97
    traffic jams, pollution, and
        accidents, 95–96
competition, 92, 104
    among bus drivers, 98
    for funding, 97
customer survey, 100
integrated land use, 93
lesson learned and future implications,
        102–105
    future implications for MENA
        region, 103–105
    lesson learned, 102–103
metro rail lines, 104
modes of transport, comparison of, 89
nitrogen dioxide, 89
non-governmental organization, 96
OECD countries, 83
operational costs, 101
Optibus, 97
poly-aromatic hydrocarbons, 89
projected 50 fastest growing urban
        areas, 86–87
public planning, 91
public-private partnerships, 82
rail transit, 89
rapid urbanization, 83–89
    environmental, health, and social
        impact, 85–89
    growth of cities and urban sprawl,
        83–85
subsidies, opponents of, 95
sulfur dioxide, 88
total suspended particles, 87, 88
traffic congestion, 85, 103
transportation and public policy,
        89–95
    bus rapid transit, 90–92
    institutional framework, 93–94
    public policy, 93
    transport economics, 94–95

urban mass rapid transportation,
89–90
urban sprawl, 83
volatile organic compounds, 88
world's most populated urban areas,
84–85
Urban sprawl, 83
Urban waste, 6
U.S. consumer price index, 126

V

Value-added tax (VAT), 124
Value chain, 170
business activity, 166
fair pricing and, 12
green lighting, 5
VAT, *see* Value-added tax
VOCs, *see* Volatile organic compounds
Volatile organic compounds (VOCs), 88

W

Waste
associated with electrical and
electronic equipment (WEEE), 6
burned, 14
disposal, 9, 162, 198
generation, 2
management, 7
most common practices, 7
responsibility, 170
packaging, 7
recycling, 6
-removal strategies, 154
seriousness of issue, 10
urban, 6
Wastewater
new compounds in, 8
treatment technologies, 196
Water

contamination, 7
filters, 1
freshwater, 9
management, 2, 8, 154, 195–197
purification, 8
resources, 4
salt, 142, 145
seawater, 197
species, 141
-table contamination, 126
travel by, 136
treatment, 8
WEEE, *see* Waste associated with
electrical and electronic
equipment
WEF, *see* World Economic Forum
WHO, *see* World Health Organization
Wind power, 3
Denmark, 44, 46
essential readings, 190
farms, 43
increasing contribution, 20
onshore, 31, 32
peak, 31
turbines, 157
Wind sector, development of, 43–46
collaboration among all industry
actors, 45–46
government coordination and
incentives, 45
grassroots support and communities,
44–45
incremental development of
technology, 46
World Economic Forum (WEF), 107
World Health Organization (WHO), 87
World Tourism Organization, 136

Z

Zero-emission vehicles, 13

For Product Safety Concerns and Information please contact our EU
representative GPSR@taylorandfrancis.com Taylor & Francis Verlag GmbH,
Kaufingerstraße 24, 80331 München, Germany

Printed and bound by CPI Group (UK) Ltd, Croydon, CR0 4YY
08/05/2025
01864510-0001